On Prevention of Significant Deterioration of Air Quality

Committee on Prevention of
Significant Deterioration
of Air Quality

Environmental Studies Board

Commission on Natural Resources

National Academy Press
Washington, D.C. 1981

NOTICE: The project that is the subject of this report was approved by the Governing Board of the National Research Council, whose members are drawn from the councils of the National Academy of Sciences, the National Academy of Engineering, and the Institute of Medicine. The members of the committee responsible for the report were chosen for their special competences and with regard for appropriate balance.

This report has been reviewed by a group other than the authors according to procedures approved by a Report Review Committee consisting of members of the National Academy of Sciences, the National Academy of Engineering, and the Institute of Medicine.

The National Research Council was established by the National Academy of Sciences in 1916 to associate the broad community of science and technology with the Academy's purposes of furthering knowledge and of advising the federal government. The Council operates in accordance with general policies determined by the Academy under the authority of its congressional charter of 1863, which establishes the Academy as a private, nonprofit, self-governing membership corporation. The Council has become the principal operating agency of both the National Academy of Sciences and the National Academy of Engineering in the conduct of their services to the government, the public, and the scientific and engineering communities. It is administered jointly by both Academies and the Institute of Medicine. The National Academy of Engineering and the Institute of Medicine were established in 1964 and 1970, respectively, under the charter of the National Academy of Sciences.

This study was funded by the U.S. Environmental Protection Agency.

Library of Congress Catalog Card Number 81-80676
International Standard Book Number 0-309-03137-0

Available from:

NATIONAL ACADEMY PRESS
2101 Constitution Ave., N.W.
Washington, D.C. 20418

Printed in the United States of America

COMMITTEE ON PREVENTION OF SIGNIFICANT DETERIORATION OF AIR QUALITY

John T. Middleton, Chairman, Washington, D.C.
Thomas D. Crocker, University of Wyoming
J. Clarence Davies, The Conservation Foundation
Charles H. Kruger, Jr., Stanford University
John H. Ludwig, Santa Barbara, California
John H. Seinfeld, California Institute of Technology
Lelan F. Sillin, Jr., Northeast Utilities Service Company
Philip Soper, University of Michigan
John Trijonis, Santa Fe Research Corporation

Myron F. Uman, Senior Staff Officer
Nancy Fairfax, Administrative Assistant

ENVIRONMENTAL STUDIES BOARD

COMMISSION ON NATURAL RESOURCES

Robert M. White, Chairman, University Corporation for Atmospheric Research
Timothy Atkeson, Steptoe & Johnson
Stanley I. Auerbach, Oak Ridge National Laboratory
Norman A. Copeland, E.I. du Pont de Nemours and Company, Inc., retired
George K. Davis, University of Florida, retired
Edward D. Goldberg, Scripps Institution of Oceanography
Charles J. Mankin, Oklahoma Geological Survey
Chester O. McCorkle, Jr., University of California, Davis
Norton Nelson, New York University Medical Center
Daniel A. Okun, University of North Carolina
David Pimentel, Cornell University
John E. Tilton, Pennsylvania State University
Alvin M. Weinberg, Oak Ridge Associated Universities
E. Bright Wilson, ex officio, Harvard University

Wallace D. Bowman, Executive Director

CONTENTS

FIGURES

TABLES

PREFACE

In accordance with provisions of the Clean Air Act Amendments of 1977 (PL 95-95) the U.S. Environmental Protection Agency asked the National Academy of Sciences to perform a study of the implementation of Part C of Title I of the Act, dealing with prevention of significant deterioration of air quality (PSD). Responding to this request, the National Research Council, which is the operating arm of the National Academy of Sciences and the National Academy of Engineering, appointed a Committee on Scientific Issues in the Prevention of Significant Deterioration of Air Quality.

The Committee, operating under the auspices of the NRC's Environmental Studies Board, conducted a two-year study. This report is the result of that work.

We conducted the study by reviewing the current literature and by consulting a large number of people who are knowledgeable in one aspect or another of the PSD provisions and their implementation (see the Appendix). We spoke with responsible officials in each of the EPA regional offices, in each of the six states with approved PSD implementation plans, and in several states whose PSD programs have not been approved by EPA; with representatives of applicants for PSD permits; with staff of EPA's Office of Air Quality Planning and Standards; and with congressional staff. We gratefully acknowledge the enthusiastic cooperation of those whom we asked for help; they gave generously of their data, analyses, and views.

Although the initial drafts of various sections were prepared by individuals, the document was extensively reviewed by the entire Committee. The report represents the collective views of the members.

We were assisted in our work by Myron F. Uman, Senior Staff Officer, and Nancy Fairfax, Administrative Assistant, of the NRC. Editorial assistance was provided by Robert C. Rooney and Catherine Iino, and the manuscript was prepared by the Manuscript Processing Unit under the capable leadership of Estelle Miller.

Finally, it is a pleasure to acknowledge the enthusiastic, dedicated, and unstinting participation of my colleagues on the Committee.

John T. Middleton, Chairman
Committee on Prevention of
Significant Deterioration
of Air Quality

SUMMARY

The Clean Air Act Amendments of 1977 (PL 95-95, 91 Stat. 731) established a national program to prevent significant deterioration of air quality (PSD) where the air is relatively clean. The amendments also mandated two independent studies of the PSD provisions of the Act, one to be performed by the National Commission on Air Quality and one by the National Academy of Sciences. This report contains the findings, conclusions, and recommendations of the Academy's study.

The objective of our study was to examine the implementation of the PSD provisions, with special emphasis on the extent to which they make sense in light of current scientific and technical knowledge. This report therefore examines what is known about the air quality related values that the PSD provisions were established to protect (Chapter 2); describes the applicability and limitations of air quality models for use in preconstruction review (Chapter 3); discusses the potential effects of the PSD provisions on siting of major emitting facilities and on industrial development (Chapter 4); addresses the availability of control technology for PSD purposes (Chapter 5); and reviews alternatives within and to the current program for achieving PSD objectives (Chapter 6). In Chapter 1 we describe the elements of the PSD program contained in the Act and elaborated in the federal regulations.

PRINCIPAL FINDINGS

The purpose of PSD is to manage economic growth in a manner consistent with concerns for good air quality.

The PSD provisions establish a series of procedures for maintaining good air quality where the air already meets national standards for protecting health and welfare. The procedures include definition of the amounts of degradation of air quality regarded as significant (called increments) in each of three categories of land in PSD areas, preconstruction review of proposed facilities for the effects of their emissions on air quality and for the appropriate levels of emission control, and means for state authorities to decide the relative importance of good air quality and industrial development in specific cases.

We have concluded that the PSD part of the Act is basically sound but is being interpreted too narrowly by the U.S. Environmental Protection Agency (EPA), state air regulatory agencies, and affected industrial concerns. These interpretations are resulting in the program being implemented with a rigidity which is contrary to Congressional intent and which, in the long run, will defeat the Act's purpose of striking a balance between economic development and values associated with clean air.

The proper interpretation of the PSD provisions is that they provide for a decision-making process through which states and federal land managers can reach determinations in specific cases about the relative importance of existing good air quality and other values. Congress recognized that there is neither adequate scientific knowledge nor a national consensus about the relevant social values to establish uniform standards dealing with visibility and other air quality related values (AQRVs). The PSD program establishes a framework within which each state can reach such a consensus according to local needs and values, except where specific national interests are at stake. But unless the states, EPA, and federal land managers adopt more flexible attitudes about the increments, the states will be unable to carry out their intended role of balancing objectives based on local values.

Explicit in the PSD provisions are a set of allowable increments of degradation of air quality beyond baseline conditions by sulfur dioxide (SO_2) and total suspended particulate mass (TSP). These increments have been treated by both permitting authorities and industry as absolute limitations on the deterioration of air quality, tantamount to supplementary air quality standards for areas of the country where the primary and secondary ambient air quality standards are already being met. As indicated by the legislative history of the Act, however, these increments were intended rather as benchmarks which, when encountered, would either trigger changes in the way in which siting and permitting decisions are to be made or signal the need to decide explicitly the relative importance of competing environmental and developmental values. The PSD provisions contain several mechanisms--as yet untested--for allowing continued industrial development when the increments have been consumed. Among these mechanisms are waivers and variances, reclassification of land, and use of operational controls on a case-by-case basis.

The PSD program is intended to be carried out by the states through mandated state implementation plans (SIPs). To date, however, only 6 states have EPA-approved plans; 13 states have been delegated authority to administer all or parts of the federal regulations. Elsewhere the PSD program is being administered under federal regulations by EPA regional offices. The EPA regional offices are less likely than the states themselves to take local considerations into account in PSD decision making.

The EPA regional offices and most states appear not to be taking seriously provisions in the Act that would ensure the greatest possible local benefit for a given decrement in air quality. The Act requires that new major stationary sources and major modifications to existing sources be subject to limitations in emissions based on a case-by-case determination of the best available control technology (BACT). The BACT requirement is intended to ensure that each PSD source uses as little of the air resource as practical, leaving as much of that resource as possible for subsequent use by others. We find that the current BACT provisions, including the requirement of case-by-case determinations, to be both appropriate and necessary to achieve the objectives of the Act.

Determinations of BACT are supposed to take into consideration energy consumption, environmental consequences, and economic impacts, but authorities have generally permitted the least-stringent emission limitations allowed under the Act--especially in cases where determinations of BACT have been made by EPA regional offices, which may understandably be reluctant to go beyond minimum requirements when acting for the states. Adopting the least stringent emission controls is usually justified as requiring the least cost to the operator. Permitting authorities have rarely taken into account other economic impacts, such as benefits to local economies of more stringent limitations which, although more costly to the operators, might allow more firms to locate in the vicinity. Present economies may be short-lived. The air resource is finite; the less each operator uses of this resource, the more is left for use by others.

The current system of increments combined with case-by-case review of emissions limitations and modified in ways described in the report could be an effective means for managing the air resource while allowing for industrial development and for protecting the national interest in good air quality in special places.

To realize the intended flexibility and local control of PSD decisions, there is much scientific work to be done. Of singular importance, means need to be developed for judging the effects of changes in air quality on AQRVs. If, as the Act intends, decisions on siting of new facilities and economic development that affect certain areas of the nation are to be made consonant with AQRVs, scientific understanding of the relationships between emissions and AQRVs must be improved. Guidance is also needed on how these consequences are to be evaluated and weighed in the PSD process.

Additional findings, conclusions, and recommendations of our study are summarized below.

AIR QUALITY RELATED VALUES

The importance of various AQRVs differs from place to place; the consequences for AQRVs of adding small concentrations of pollutants to the ambient air depend on existing air quality. Visibility, odor, vitality of flora and fauna, integrity of geological features and cultural resources, and the acidity of precipitation are several among many AQRVs. Some of these values are linked. For example, acidity of both wet and dry deposition can have an adverse effect on cultural resources such as monuments and buildings.

While the relative importance of specific values varies from place to place, there is a general consensus that visibility is the most important value, particularly in the West. Visibility is the only AQRV specifically addressed in the Clean Air Act as amended. In the East, nevertheless, the acidity of precipitation and potential regional climate modification appear to rival visibility as the AQRV of greatest concern.

There is good scientific understanding of the effects of particles and gases on the optical properties of the atmosphere; those effects depend strongly on background concentrations. The relationships between incremental changes in atmospheric concentrations of pollutants and degradation of other AQRVs are not always clearly understood. Even less is understood about human sensory perception of changes in AQRVs.

The pollutants for which the Act defines increments of degradation, SO_2 and TSP, are not in all cases directly related to the AQRVs of greatest concern. Sulfur dioxide and TSP are two pollutants originally regulated under the Act to protect human health and welfare. But the relationships between ambient concentrations of these pollutants and most AQRVs--visibility in particular--are only indirect. The PSD program would be on firmer scientific ground if increments were established for pollutants more directly related to AQRVs. Specifically, we recommend that an increment be established for fine particulate matter, particles with diameters of less than about 2.5 micrometers.

Although the so-called criteria pollutants do not always bear direct relationships to AQRVs, control of emissions of these pollutants may nevertheless be beneficial for preventing significant deterioration of air quality. Sulfur dioxide, for example, is a colorless gas, but some emissions of this gas are transformed by atmospheric processes into fine sulfate particulate matter, the presence of which has an adverse effect on visibility. Controlling emissions of sulfur dioxide and the oxides of nitrogen limits atmospheric formation of sulfates, nitrogen dioxide, and nitrates, which have direct adverse effects on several AQRVs.

The current PSD program may be of limited usefulness for addressing regional air quality problems. Some kinds of degradation of air quality are directly related to emissions from a specific source; this is the case, for example, with plume blight and odors. Other problems, such as general regional hazes or acid precipitation, are caused by pollutants that may be transported long distances and are likely to be mixtures of emissions from many sources.

Two fundamental reasons why the PSD program may be ineffective for such large-scale, multistate air quality problems are that modeling cannot separate the influences of individual sources hundreds of kilometers away and that there is no effective administrative procedure in the Act for dealing with air quality problems that span many states. Another reason PSD is not well suited to the problems of large-scale regional haze and acid precipitation is that PSD focuses mainly on emissions from new sources. For the next one or two decades, the main sources of SO_2, which contributes to regional haze and acid precipitation through its transformation products, will be poorly controlled existing sources rather than well-controlled new sources. An alternative approach to regional air quality problems is establishment of limitations on total regional emissions.

Except for programs to monitor visibility, the monitoring programs required by the PSD regulations do not fully support the objectives of the PSD provisions. The monitoring programs required under the PSD regulations focus on the criteria pollutants and, to some extent, on visibility. The greater part of this monitoring is aimed at determining background concentrations of SO_2 and TSP at or near sites of proposed facilities; very little monitoring of the status of AQRVs themselves, other than visibility, is being done. Because of the lack of monitoring data for and the poor understanding of many AQRVs, it may be very difficult to ascertain if and when these values are threatened.

Preconstruction and postconstruction monitoring may be useful in some cases for validating air quality models. Uniform requirements that all prospective new sources in PSD areas conduct preconstruction monitoring are unnecessary. Amendment of the PSD provisions to provide permitting authorities flexibility to impose preconstruction monitoring requirements would make the permitting process simpler, faster, and less expensive without reducing the effectiveness of PSD.

AIR QUALITY MODELING

Mathematical models are the only tools available for use in predicting the effects of emissions on air quality. The continued use of mathematical descriptions of the behavior of pollutants emitted into the atmosphere is essential to the implementation of the PSD program. Models are used to predict changes in atmospheric concentrations of pollutants as part of the preconstruction review of proposed major facilities. A limited number of models are available for predicting the effects of emissions on visibility. Improved models incorporating point and distributed sources are also required for keeping track of the status of increments.

Models currently being used in the PSD process are inappropriate for many of the applications intended. Conservative screening models are being used to determine the emission limitations required to avoid exceeding increments; models designed and applicable for short-range studies are being used in situations where long-range transport is the critical feature; and steady-state models are being used to describe dynamic conditions. Use of conservative models to screen permit applications is not inappropriate as a means for reducing administrative burdens of the preconstruction review process. If, however, such models are used to determine the availability of increments or to set emission limitations for facilities whose emissions threaten to cause increments to be exceeded, predictions of consumption of increments will be unrealistic and limitations may be more stringent than necessary. Continued development of models that can be used for complex terrain and long-range transport is vital.

The results of modeling for short-term (3-hour and 24-hour) maxima are inherently uncertain because the maximum occurrence is a rare event. Modeling for the maximum value over short averaging times poses different problems than modeling for long-term average values. Different models may produce different results when predicting rare events owing to differences in treatments of meteorological variables. On the other hand, results of modeling for long-term average values with different models usually agree rather closely. Therefore, to the extent that the objectives of the PSD program can be met by concentrating on long-term averages (annual or seasonal) as opposed to short-term maxima, disputes about which models to use can be avoided.

We also find that greater confidence in the results of modeling might be achieved with a change in the criteria for compliance. If, for example, instead of the current requirement that short-term increments not be exceeded more than once per year, there were a requirement expressed statistically (for instance, that a given level could not be exceeded, say, five percent of the time), the statistical reliability of the results of modeling would be greatly enhanced. Such a change in compliance criteria need not result in altered levels of protection of air quality.

SITING AND INDUSTRIAL DEVELOPMENT

The costs associated with PSD are difficult to assess. The effects of the program depend on characteristics of specific situations, and the program allows many variations in application. Moreover, even in specific cases it may not be entirely possible to separate costs of PSD--such as incremental costs of installing and operating pollution control equipment, differential siting costs, and costs associated with delays in obtaining permits--from other costs.

The purpose of PSD is to bring considerations of air quality into decision making on the siting of new facilities; therefore it should not be surprising that implementation of PSD regulations can affect siting. Finding suitable sites for major new facilities normally involves satisfying a wide variety of constraints, such as proximity to resources, transportation systems, and markets. PSD adds such considerations as proximity to Class I areas and to other sources subject to the PSD regulations, availability of increments, and effects on AQRVs.

Drawing general conclusions about the distances that must separate certain types of sources from Class I areas (where existing good air quality is most highly valued and which therefore require the highest level of protection) or from one another is inappropriate, because so much depends on industrial processes, control technologies, terrain, and prevailing meteorological conditions. Nonetheless, it is clear that the existence of Class I areas may constrain the siting of new major facilities in the West and the proximity to other sources subject to PSD may constrain siting in some Class II areas in the East.

The options provided in the Act for allowing industrial development despite increment consumption have yet to be tested. There has thus been insufficient experience to indicate how successful the processes for decision making will be in allowing continued development when increments have been consumed.

THE BACT REQUIREMENT

Technologies are being developed to control direct emissions of pollutants regulated under the Act; available technologies for controlling emissions that adversely affect AQRVs are adequate for many but not all PSD purposes. Technology is available for controlling direct emissions of SO_2 and TSP. With anticipated increases in coal combustion, however, improved measures for controlling emissions of the oxides of nitrogen are needed. For the most part, advanced technologies for controlling TSP emissions effectively control direct emissions of fine particles, although removal efficiencies for fine particles are less than those for total particulate mass. Greater attention may have to be paid to controlling fine-particle emissions where these particles are significant in comparison with fine particles formed in the atmosphere from precursor gases.

Existing sources of emissions may contribute significantly to the degraded state of visibility that currently exists in many areas. In these cases retrofit technology is required by the Act to restore visibility. Existing sources also contribute to the current state of acidity of precipitation. Efforts aimed at reducing both regional haze and acid rain problems will probably not be successful for several decades without control of existing sources.

To date, determinations of best available control technology for new sources generally have established emission limitations more stringent than the minimum requirement only when emissions have directly threatened to cause increments to be exceeded. Permitting authorities have been lenient in applying the definition of BACT; sources have generally been constrained only by new source performance standards or their equivalents. The number of permits that can be issued before increments are consumed has thus been artificially constrained. Some authorities, including some EPA regional offices, have determined BACT for classes of sources rather than on a case-by-case basis as the law requires. Many BACT determinations have not included assessments of alternative technologies or taken into account the environmental, energy, and economic consequences--other than direct costs to operators--of proposed controls.

State agencies need help in carrying out their responsibilities under the PSD program, both in determining BACT on a case-by-case basis and in modeling the effects of emissions on air quality. EPA maintains a clearinghouse for exchanging information on BACT determinations and publishes modeling guidelines; both efforts need continual improvement and updating. Continuing guidance on the applicability of retrofit controls is also needed.

To the extent that emission limitations of BACT are more stringent than those of NSPS, determinations of BACT may incorporate intermittent controls. Using intermittent controls, such as reduced operations or fuel switching, may be attractive where emissions from a facility already using the best available continuous control technology still threaten to cause increments in a distant Class I area to be exceeded under infrequent meterological conditions. A critical factor in deciding on intermittent controls in each case is the acceptable level of confidence with which specific meteorological conditions--identified by modeling as those under which additional controls are required--can be anticipated or predicted. Another consideration is that the use of intermittent controls could protect AQRVs in a Class I area from plume blight, but would increase the overall regional emissions burden from what it would be if the facility were not built.

IMPLEMENTATION AND ALTERNATIVES

In the implementation of PSD, allocation of the increments has been on a first-come, first-served basis. Permission to use all or part of the scarce air resource is being granted solely on the basis of the applicant's place in the queue. Other administrative criteria, such as type of facility or number of local jobs created, have thus far not been employed. No allocation system that employs economic incentives is currently in use. Neither the Act nor federal regulations constrain the selection of allocation systems.

Alternatives to the current regulatory system for implementing PSD have been proposed, but in practical application, none appears to offer clear advantages over the regulatory system. Among the alternative schemes for allocating access to the air resource most often mentioned are schemes employing economic incentives rather than administrative regulations; emission taxes and marketable emission permits are two prominent examples. Studies of the advantages and applicability of these types of systems have been conducted primarily for situations that obtain in nonattainment areas and therefore may be of limited use in deciding the applicability of economic incentives to PSD.

For most economic systems to work, a substantial number of sources are required. PSD areas are frequently characterized by few sources, with some sources having the potential of consuming very substantial portions of the available increments. Thus it is not obvious that the desirable properties of competition inherent in economic allocation schemes would consistently emerge in PSD areas.

1
INTRODUCTION

One purpose of Title I of the Clean Air Act (42 USC 7401 et seq.) is "to protect and enhance the quality of the Nation's air resources so as to promote the public health and welfare and the productive capacity of its population." The interpretation of this phrase has been the subject of public debate and of a suit, Sierra Club et al. v. Ruckelshaus.[1] The court found that the Act requires an air quality program aiming at not only improving air quality in places where the air is relatively dirty (judged against national standards for public health and welfare) but also preventing serious degradation of air quality where the air is relatively clean.

In response to the decision in the Sierra Club suit, the U.S. Environmental Protection Agency (EPA) issued regulations on December 5, 1974, requiring each state to adopt a program to prevent significant deterioration of air quality. Congress affirmed the concept of the prevention of significant deterioration (PSD) by adding Part C to Title I of the Act (42 USC 7470) in the Clean Air Act Amendments of 1977 (PL 95-95, 91 Stat. 685).

Subsequently, EPA revised its PSD regulations on June 19, 1978. The new regulations were the subject of another law suit, Alabama Power v. Costle (606 F.2d 1068 [D.C. Cir. 1979]), challenging several of the agency's

---

[1]Sierra Club et al. v. Ruckelshaus [344 F. Supp. 253 (D.D.C.)] was affirmed on appeal [Env. Rep. Cas. 1815 (D.C. Cir. 1972)] and again affirmed by an equally divided Supreme Court [412 U.S. 541 (1973)].

interpretations of the Act. Following adjudication of Alabama Power vs. Costle, EPA revised its regulations again (45 Fed. Reg. 52676, August 7, 1980).

In this chapter, we shall describe the current PSD program as established in the Act and elaborated in EPA's regulations. We shall also discuss alternative ways of understanding of the PSD provisions. This chapter lays the groundwork for the more detailed analyses of important aspects of PSD presented in the rest of this report.

PURPOSES OF THE NAAQS AND OF PSD

Preventing the degradation of air quality where the air is relatively clean and improving air quality where the air is unacceptably dirty imply the existence of measures of acceptable air quality. Section 109 of the Clean Air Act requires the administrator of EPA to define these measures by promulgating national ambient air quality standards (NAAQS) for pollutants that cause or contribute to air pollution that may endanger public health and welfare. The Act defines two types of standards. The primary NAAQS are intended to protect public health, allowing for an adequate margin of safety. The secondary NAAQS are intended to protect the public welfare from known or anticipated adverse effects. Under Section 110, the states are required to adopt plans to implement, maintain, and enforce the NAAQS; each state implementation plan (SIP) must be approved by the administrator of EPA.

The PSD provisions of the Act, Sections 160 through 169A, apply to areas of the country where ambient levels of those pollutants for which NAAQS have been promulgated are below the standards--that is, where the air is relatively clean. (Sections 171 through 178 of the Act deal with nonattainment areas, where ambient levels are higher than allowed by the NAAQS.) The Act specifies five purposes for the PSD provisions:

- to protect public health and welfare from any actual or potential adverse effect that may reasonably be anticipated even if the NAAQS are met;
- to preserve, protect, and enhance air quality in areas of special national or regional interest;

- to ensure that economic growth will occur in a manner consistent with the preservation of existing clean air resources;
- to ensure that emissions from any source in any state will not interfere with any portion of the applicable implementation plan to prevent significant deterioration for any other state; and
- to ensure that any decision to permit increased air pollution in any area to which the PSD program applies is made only after careful evaluation of all the consequences and after adequate procedural opportunities for informed public participation in the decision-making process.

The NAAQS are uniform minimum national standards for air quality. The PSD provisions provide for additional protection in areas where existing air quality is better than the minimum standards require. Where existing good air quality is deemed to be of national importance, the states are required to provide additional protection to prevent significant deterioration of that good air quality. The Act establishes procedures by which state authorities may decide, within certain constraints and conditions, what levels of protection of air quality in other areas are appropriate for the needs of their states.

THE PSD PROVISIONS

State Implementation Plans

The provisions of the Clean Air Act are intended to be carried out by state authorities; under Section 110, the states are required to prepare implementation plans for approval by the administrator of EPA. Until EPA approves the PSD portion of a plan for a state, the federal PSD regulations are administered in that state by EPA through one of its ten regional offices. In some circumstances, an EPA regional office may delegate to a state responsibility to perform part or all of PSD reviews; in this situation, a state implements the federal regulations but may also adopt and enforce more stringent regulations. Table 1.1 shows the status of state PSD programs as of November 1, 1980. Several of the states for which EPA currently administers the

TABLE 1.1. Status of state PSD programs as of November 1, 1980.

EPA Region	Approved PSD SIP	Full or Partial Delegation of PSD Review to State	PSD Administered Completely by EPA
I	Maine, Vermont		Connecticut, Massachusetts, New Hampshire, Rhode Island
II			New York, New Jersey, Puerto Rico, Virgin Islands
III			Delaware, District of Columbia, Maryland, Pennsylvania, Virginia, West Virginia
IV	Georgia, Tennessee	Alabama, Florida, Kentucky, Mississippi, North Carolina, South Carolina	
V		Illinois, Indiana, Michigan, Minnesota, Ohio, Wisconsin	
VI			Arkansas, Louisiana, New Mexico, Oklahoma, Texas
VII		Nebraska	Iowa, Kansas, Missouri
VIII	North Dakota, Wyoming		Colorado, Montana, South Dakota, Utah
IX			Arizona, California, Guam, Hawaii, Nevada
X			Alaska, Idaho, Oregon, Washington

federal PSD program also have their own programs, which are being implemented even though they have not been approved by EPA; in these few cases, proposed new sources of air pollution are subject to separate state and federal PSD reviews.

There appear to be two major reasons why more states have not prepared SIPs or accepted delegation of the federal program. One reason is that running a PSD program requires considerable financial resources and highly trained scientific personnel. A PSD program must compete with other state priorities for limited resources. The other is that the frequent changes in the details of the federal regulations have created uncertainties about the final form of the program to which the states must respond.

To date, SIPs have had to be revised three times to meet PSD requirements: initially to implement the regulations promulgated following the decision in Sierra Club v. Ruckelshaus, then in response to changes in regulations required by the passage of the Clean Air Act Amendments of 1977, and once again in response to changes in the federal regulations made after the decision in Alabama Power v. Costle. Two states, Georgia and North Dakota, submitted SIPs for approval in response to the original regulations; the other states with approved PSD SIPs responded to the 1977 amendments. A number of states are in the process of preparing their plans, having waited until Alabama Power v. Costle was resolved. New litigation over the federal regulations promulgated in response to the Alabama Power decision may further delay state implementation of the program.

Classification of Areas

The areas of the country covered by the PSD provisions are divided into three classes. Class I areas are those where exceptionally good air quality is most highly valued. Air quality values are less highly prized in Class II and Class III areas. Different degrees of degradation of air quality are deemed acceptable in the three classes of land, with the least deterioration in Class I and the most in Class III.

In the Clean Air Act Amendments of 1977, Congress specified the initial classification of lands for PSD purposes. Certain lands, where existing good air quality is deemed to be of national importance, were designated Class I and may not be reclassified. These mandatory Class I areas include all international parks, national wilderness areas larger than 5,000 acres, national memorial parks larger than 5,000 acres, and national parks larger than 6,000 acres that were in existence when the amendments were passed. Figure 1.1 shows the mandatory Class I areas. In addition, Congress provided that any other areas that had been redesignated Class I under EPA's previous PSD regulations would remain Class I; these areas, however, are subject to reclassification. All other areas to which the PSD provisions apply were initially designated Class II.

States and Indian tribes may reclassify lands under their jurisdiction except for the mandatory Class I areas. Procedures for reclassification are specified in Section 164 of the Act. States are required to give public notice, to notify federal officials if the land under consideration for redesignation includes federal lands, to issue analyses of the effects of redesignation, and to hold public hearings. Redesignation of non-Indian

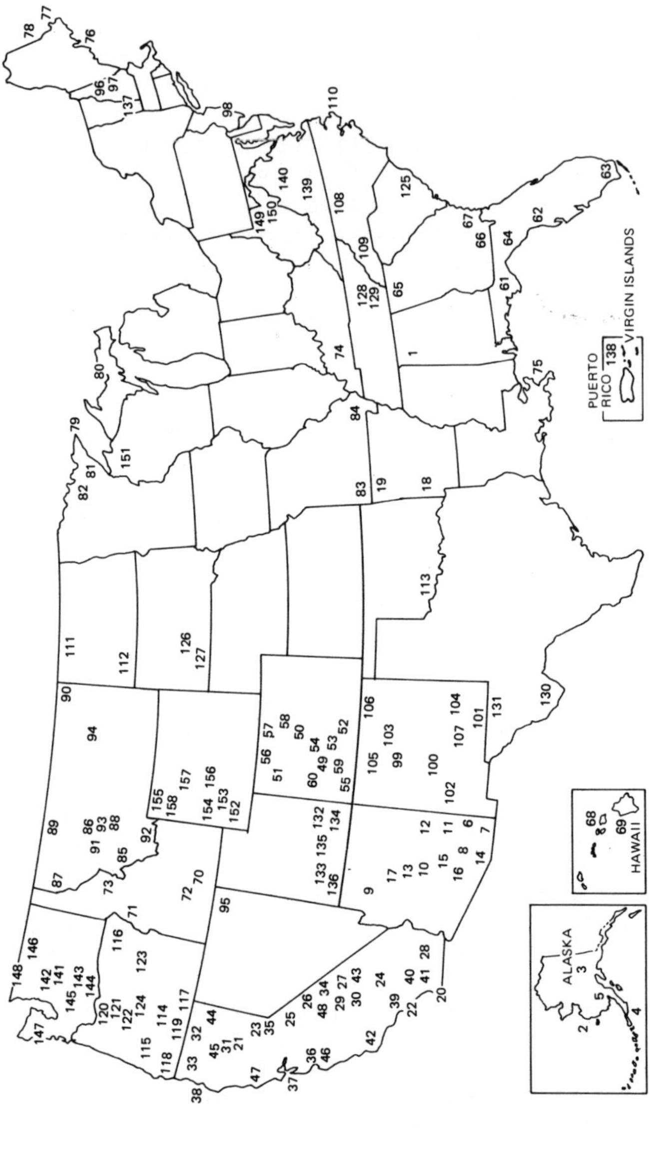
ALASKA
HAWAII
PUERTO RICO
VIRGIN ISLANDS

Alabama
- Sipsey Wild 1

Alaska
- Bering Sea Wild 2
- Mount McKinley NP 3
- Simeonof Wild 4
- Tuxedni Wild 5

Arizona
- Chiricahua NM Wild 6
- Chiricahua Wild 7
- Galiuro Wild 8
- Grand Canyon NP 9
- Mazatzal Wild 10
- Mount Baldy Wild 11
- Petrified Forest NP 12
- Pine Mountain Wild 13
- Saguaro Wild 14
- Sierra Ancha Wild 15
- Superstition Wild 16
- Sycamore Canyon Wild 17

Arkansas
- Caney Creek Wild 18
- Upper Buffalo Wild 19

California
- Agua Tibia Wild 20
- Caribou Wild 21
- Cucamonga Wild 22
- Desolation Wild 23
- Dome Land Wild 24
- Emigrant Wild 25
- Hoover Wild 26
- John Muir Wild 27
- Joshua Tree Wild 28
- Kaiser Wild 29
- Kings Canyon NP 30
- Lassen Volcanic NP 31
- Lava Beds Wild 32
- Marble Mountain Wild 33
- Minarets Wild 34
- Mokelumne Wild 35
- Pinnacles Wild 36
- Point Reyes Wild 37
- Redwood NP 38
- San Gabriel Wild 39
- San Gorgonio Wild 40
- San Jacinto Wild 41
- San Rafael Wild 42
- Sequoia NP 43
- South Warner Wild 44
- Thousand Lakes Wild 45
- Ventana Wild 46
- Yolla-Bolly-Middle-Eel Wild 47
- Yosemite NP 48

Colorado
- Black Canyon of the Gunnison Wild 49
- Eagles Nest Wild 50
- Flat Tops Wild 51
- Great Sand Dunes Wild 52
- La Garita Wild 53
- Maroon Bells-Snowmass Wild 54
- Mesa Verde NP 55
- Mount Zirkel Wild 56
- Rawah Wild 57
- Rocky Mountain NP 58
- Weminuche Wild 59
- West Elk Wild 60

Florida
- Bradwell Bay Wild 61*
- Chassahowitzka Wild 62
- Everglades NP 63
- St. Marks Wild 64

Georgia
- Cohotta Wild 65
- Okefenokee Wild 66
- Wolf Island Wild 67

Hawaii
- Haleakala NP 68
- Hawaii Volcanoes 69

Idaho
- Craters of the Moon Wild 70
- Hells Canyon Wild 71
- Sawtooth Wild 72
- Selway-Bitterroot Wild 73

Kentucky
- Mammoth Cave NP 74

Louisiana
- Breton Wild 75

Maine
- Acadia NP 76
- Moosehorn Wild 77
- Roosevelt Campobello IP 78

Michigan
- Isle Royale NP 79
- Seney Wild 80

Minnesota
- Boundary Waters Canoe Area Wild 81
- Voyageurs NP 82

Missouri
- Hercules-Glades Wild 83
- Mingo Wild 84

Montana†
- Anaconda-Pintlar Wild 85
- Bob Marshall Wild 86
- Cabinet Mountains Wild 87
- Gates of the Mountain Wild 88
- Glacier NP 89
- Medicine Lake Wild 90
- Mission Mountain Wild 91
- Red Rock Lakes Wild 92
- Scapegoat Wild 93
- U.L. Bend Wild 94

Nevada
- Jarbidge Wild 95

New Hampshire
- Great Gulf Wild 96
- Presidential Range-Dry River Wild 97

New Jersey
- Brigantine Wild 98

New Mexico
- Bandelier Wild 99
- Bosque del Apache Wild 100
- Carlsbad Caverns NP 101
- Gila Wild 102
- Pecos Wild 103
- Salt Creek Wild 104
- San Pedro Parks Wild 105
- Wheeler Peak Wild 106
- White Mountain Wild 107

North Carolina
- Linville Gorge Wild 108
- Shining Rock Wild 109
- Swanquater Wild 110

North Dakota
- Lostwood Wild 111
- Theodore Roosevelt NP 112

Oklahoma
- Wichita Mountains Wild 113

Oregon
- Crater Lake NP 114
- Diamond Peak Wild 115
- Eagle Cap Wild 116
- Gearhart Mountain Wild 117
- Kalmiopsis Wild 118
- Mountain Lakes Wild 119
- Mount Hood Wild 120
- Mount Jefferson Wild 121
- Mount Washington Wild 122
- Strawberry Mountain Wild 123
- Three Sisters Wild 124

South Carolina
- Cape Romain Wild 125

South Dakota
- Badlands Wild 126
- Wind Cave NP 127

Tennessee
- Great Smoky Mountains NP 128
- Joyce Kilmer-Slickrock Wild 129

Texas
- Big Bend NP 130
- Guadalupe Mountains NP 131

Utah
- Arches NP 132
- Bryce Canyon 133
- Canyonlands NP 134
- Capitol Reef NP 135
- Zion NP 136

Vermont
- Lyle Brook Wild 137

Virgin Islands
- Virgin Islands NP 138

Virginia
- James River Face Wild 139
- Shenandoah NP 140

Washington
- Alpine Lakes Wild 141
- Glacier Peak Wild 142
- Goat Rocks Wild 143
- Mount Adams Wild 144
- Mount Rainier NP 145
- North Cascades NP 146
- Olympic NP 147
- Pasayten Wild 148

West Virginia
- Dolly Sods Wild 149
- Otter Creek Wild 150

Wisconsin
- Rainbow Lake Wild 151*

Wyoming
- Bridger Wild 152
- Fitzpatrick Wild 153
- Grand Teton NP 154
- North Absaroka Wild 155
- Teton Wild 156
- Washakie Wild 157
- Yellowstone NP 158

*Area where visibility is not recommended as an important value.
†In addition, the Northern Cheyenne Indian Reservation has been designated Class I by the tribal authority.

IP = International Park
Wild = Wilderness Area
NP = National Park

FIGURE 1.1. Federally mandated Class I areas.

land under state jurisdiction to Class III must have the specific approval of the governor of the state, who must consult with both the state legislature and general purpose units of affected local governments, all of which must concur with the reclassification. National monuments, national primitive areas, national preserves, national recreation areas, national wild and scenic rivers, national wildlife refuges, and national lakeshores and seashores larger than 10,000 acres as well as national parks and national wilderness areas larger than 10,000 acres that are established after the passage of the 1977 amendments cannot be redesignated Class III.

Defining Significant Deterioration

The Act defines significant deterioration due to the presence of sulfur dioxide (SO_2) and total suspended particulate matter (TSP) by setting maximum allowable increases over baseline concentrations (increments) for these pollutants for each of the three classes of land, subject to the restriction that the allowed increases may not result in concentrations that exceed either the primary or the secondary NAAQS. Increments for SO_2 and TSP--called the Set I pollutants--are defined for both a long-term (annual) average concentration and maximum concentrations over short periods of time (Table 1.2). The 24-hour and 3-hour maxima and standards may be exceeded once per year.

As a result of the decision in Alabama Power v. Costle, the baseline concentration includes (a) the ambient concentration resulting from actual emissions from existing sources at the time of the first

TABLE 1.2. NAAQS for SO_2 and TSP and maximum allowable increases in concentrations over the baseline in PSD areas ($\mu g/m^3$).

	Maximum Allowable Increases			NAAQS	
Pollutant	Class I	Class II	Class III	Primary	Secondary
Particulate Matter					
Annual geometric mean	5	19	37	75	60
24-hour maximum*	10	37	75	260	150
Sulfur Dioxide					
Annual arithmetic mean	2	20	40	80	–
24-hour maximum*	5	91	182	365	–
3-hour maximum*	25	512	700	–	1,300

*Short-term maxima may be exceeded no more than once per year.

application for a permit to construct a major emitting facility in a PSD area, and (b) the projected emissions from major stationary sources that began construction before January 6, 1975 but had not begun operation by the time of the first permit application. Baseline conditions also account for natural phenomena that affect air quality.

Contributions to ambient concentrations of SO_2 and TSP from certain sources are not to be taken into account in determining the extent to which the increments have been consumed. Exemptions are provided for emissions from plants that are required by law to convert from oil or natural gas to coal and from new sources located outside the United States. In addition, concentrations of TSP resulting from temporary construction activities are not counted against the increments. The exemption of these sources is one reason why simply measuring changes in ambient concentrations over a period of time cannot be the basis of a procedure to determine compliance.

The 1977 amendments also require EPA to devise means for preventing significant deterioration of air quality from other pollutants regulated under the Act. The means need not include a system of increments. EPA is developing a PSD program for carbon monoxide, hydrocarbons, lead, nitrogen oxides, and ozone (called the Set II pollutants). EPA plans to propose regulations for the Set II program late in 1981 and to promulgate final regulations in 1982.

While the SO_2 and TSP increments are uniformly applicable in each of the three classes of land, the Act also contains provisions for determining on a case-by-case basis the extent to which a proposed deterioration in a mandatory Class I area is significant. A proposed degradation of air quality is to be judged taking into account the air quality related values (including visibility) that are important to the specific Class I area, whether or not the increments are exceeded (see the section on the AQRV Test for Class I Areas below).

Managing the Air Resource

Establishing restrictions on allowable degradation of air quality makes explicit the notion that clean air is a finite resource. Part of a state's implementation plan is its system for managing access to that resource--that is, for allocating the consumption of the available increments among sources of emissions. A number of allocation schemes have been suggested; the Act makes no restrictions on the systems chosen by the states.

Schemes for allocating increments fall into two general categories: administrative mechanisms and market processes. Under an administrative mechanism, the permitting authority establishes criteria for allocation and judges each application by those criteria. The most common administrative practice is to allow potential emitters to consume increments on a first-come, first-served basis. In allocation schemes based on market processes, the state either establishes an economic incentive (e.g., emission charges) or creates an economic market (e.g., marketable emission permits) in which access to the air resource becomes a marketable commodity.

As of October 1, 1980, all existing allocation schemes were based on the principle first-come, first-served, with additional restrictions varying from state to state. EPA uses a first-come, first-served system when it exercises PSD review authority in the absence of an approved SIP. The amount of increment that is allocated to each applicant is formally a matter for negotiation between the applicant and the permitting authority. In practice, some agencies will not permit an applicant to consume more than a certain fixed percentage of the portion of increment remaining. Two purposes have been cited for this restriction: to ensure that some increment remains for future applicants and to account for effects of emissions from nonmajor sources that are otherwise neglected. A system of marketable permits has been proposed and is under study in at least one state, Maryland.

To issue a permit to construct a major emitting facility in a PSD area under an administrative allocation scheme, the permitting authority must determine, on the basis of demonstrations by the applicant, that the emissions will not cause or contribute to ambient concentrations of air pollutants that exceed the PSD increments or the NAAQS. The permitting process must also include analyses of the air pollution control technology and any air quality effects that would result from growth associated with a proposed facility. A public hearing must be held before a permit can be issued.

It is up to the permitting authority to keep track of changes in emissions and consumption of the increments.

Preconstruction Review

To assess the potential consequences of emissions from new sources for ambient air quality and the consumption of increments, proposed sources of pollution must be reviewed before construction. However, not all sources are subject to preconstruction review for PSD purposes. Only major stationary sources or major modifications to existing stationary sources in areas in which the PSD provisions apply are subject to review. As a result of the decision in Alabama Power v. Costle, a major source located in a nonattainment area is not subject to PSD review even if emissions from the source adversely affect air quality in a PSD area, unless the affected PSD area and the source are in different states.

There are two classes of major stationary sources. In one class are all stationary sources that emit or have the potential to emit 250 or more tons per year of any pollutant regulated under the Act. In the other class are any stationary sources from a list of 28 types of industrial facilities that emit or have the potential to emit 100 tons or more per year of any regulated pollutant. Table 1.3 lists the types of facilities in the second category.

In EPA's earlier regulations, the potential for a source to emit pollutants was to be judged without taking into account any air pollution control equipment. The decision in Alabama Power overturned this interpretation, and thus potential to emit now means the maximum capacity of a source to emit pollutants under its actual physical and operational design, which includes any air pollution control equipment. The decision means that some new sources that otherwise would have been subject to preconstruction review will now be treated as nonmajor sources, not subject to preconstruction review but included in the inventory of sources whose emissions contribute to the consumption of increments.

Major modifications of existing sources are defined as physical or operational changes that result in a significant net increase in emissions of any pollutant regulated under the Act. Regulations specify what a significant increase is for each pollutant. For example, a significant increase in the emission rate of SO_2 is 40 tons per year; of asbestos, 0.007 tons per year (45 Fed. Reg. 52732, August 7, 1980).

Stationary point sources that are not large enough to be classified as major sources and distributed sources such as urban growth associated with major sources,

TABLE 1.3. Categories of major stationary PSD sources emitting 100 tons or more per year.

Fossil fuel fired steam electric plants of more than 250 million Btu per hour heat input
Coal cleaning plants (with thermal dryers)
Kraft pulp mills
Portland cement plants
Primary zinc smelters
Iron and steel mill plants
Primary aluminum or reduction plants
Primary copper smelters
Municipal incinerators capable of charging more than 250 tons of refuse per day
Hydrofluoric acid plants
Sulfuric acid plants
Nitric acid plants
Petroleum refineries
Lime plants
Phosphate rock processing plants
Coke oven batteries
Sulfur recovery plants
Carbon black plants (furnace process)
Primary lead smelters
Fuel conversion plants
Sintering plants
Secondary metal production plants
Chemical process plants
Fossil fuel boilers (or combinations thereof) totaling more than 250 million Btu per hour heat input
Petroleum storage and transfer units with a total storage capacity exceeding 300 thousand barrels
Taconite ore processing plants
Glass fiber processing plants
Charcoal production plants

including emissions from mobile sources that accompany this growth, are not subject to preconstruction review; however, changes in air quality due to emissions from such sources are to be counted against the increments. It is up to the permitting authorities to maintain inventories of the nonmajor sources as well as the major sources and to assess the effects of emissions from both major and nonmajor sources.

Modeling the Consequences of Emissions

The consequences of emissions on ambient concentrations of pollutants and on air quality related values must be predicted before facilities are built if deterioration of air quality is to be avoided. Air quality models--which are simply mathematical representations of the physical and chemical processes to which emissions are subjected in the atmosphere--are the only tools available for this purpose. Implementing the PSD provisions therefore depends strongly on modeling capabilities. To ensure a reasonably uniform treatment of meteorological data and atmospheric phenomena in models used for PSD purposes, EPA has issued guidelines on modeling (U.S. EPA 1978, 1980c). EPA also has made available a number of models, which are in widespread use in the PSD program (see Chapter 3).

Emission Limitations for Major PSD Sources

Major stationary sources in PSD areas are required to limit emissions of each pollutant subject to regulation under the Act by using the best available control technology (BACT). The permitting authority is to determine what control technology is best and available for each specific case, taking into account energy, environmental, and economic impacts and other costs. If a source qualifies as a major one for any regulated pollutant, BACT must be applied for all regulated pollutants. The results of modeling are often important determinants in the determination of emission limitations.

Where new source performance standards (NSPS) exist for a type of source, the BACT emission limitation may not be less stringent than the NSPS limitation. The NSPS for a type of source are national standards for all new stationary sources in that class, reflecting the degree of emission limitation and percentage reduction in emissions achievable through application of the best technological system of continuous emission reduction that the administrator of EPA deems to have been adequately demonstrated. The determination of NSPS is to take into account costs of reducing emissions, health and environmental consequences of emission reductions other than those associated with air quality, and energy requirements of the control process. As of October 1980, NSPS have been promulgated for pollutants emitted by 18 types of industrial facilities of the 28 classes of major PSD sources (Env. Rep. ER 121:1501).

The BACT requirement for PSD may also be contrasted with requirements of the Act for emission limitations in nonattainment areas. Under Section 173, new facilities proposing to locate in nonattainment areas must be limited to the lowest achievable emission rate (LAER), which is defined as either the most stringent emission rate achieved in practice by a class or category of sources or the most stringent limitation in any SIP, whichever is more stringent. The determination of LAER is therefore not made on a case-by-case basis and does not take into consideration energy, environmental, or economic impacts and other costs.

Thus, the BACT requirement is at least as stringent as an applicable NSPS, but in general is not as stringent as the LAER requirement to which the new source would be subject in a nonattainment area. In some circumstances, of course, BACT could be as stringent as LAER. Vermont, for example, has adopted a PSD SIP that requires the same emission limitation in both nonattainment and PSD areas.

Nonmajor stationary sources are exempted from preconstruction review and are not required to use BACT. These sources are, however, subject to applicable NSPS.

The AQRV Test for Class I Areas

To represent the national interest in Class I areas that encompass federal lands, federal officials have been given special roles in the preconstruction review of proposed facilities that may affect those lands. Under the Act, the federal land manager and the federal official who is directly responsible for managing federal land in a Class I area have an affirmative responsibility to protect AQRVs in that land. One way of discharging that responsibility is to determine whether a proposed major source will adversely affect those values. The federal land manager is the secretary of the department with authority over the federal land in question. Other than visibility, the Act does not specify what AQRVs are (see Chapter 2).

Even if Class I increments have not been consumed and a proposed new source would not cause or contribute to concentrations of pollutants that exceed the increments, a construction permit cannot be issued if the federal land manager demonstrates to the satisfaction of the state that the proposed facility would nonetheless have adverse consequences for AQRV on the federal lands in question.

Conversely, when Class I increments have been consumed or when a proposed source would cause or contribute to concentrations of pollutants that do exceed the increments, a permit may still be issued if the applicant can demonstrate to the satisfaction of the federal land manager that the emissions from the facility will not have adverse consequences for AQRVs on federal lands in the Class I area. The burden of proof in these circumstances is on the applicant. When such waivers of the Class I increments are granted, the Act specifies new maximum allowable increases over ambient concentrations (Table 1.4), which are the same as the increments for Class II areas except for the 3-hour maximum for SO_2. The allowable increases of Table 1.4 may not be exceeded.

The Act provides another procedure for obtaining relief from the increments. The applicant must demonstrate to the satisfaction of the governor that the facility cannot be built because of the 24-hour and 3-hour SO_2 increments and, in the case of a federal mandatory Class I area, that a variance will not adversely affect AQRVs. If the governor is convinced and if the federal land manager concurs, the state may grant a variance of the SO_2 increment and issue a permit subject to the maximum allowable increases listed in Table 1.5. If the federal land manager does not concur with the findings of the governor, the recommendations of the governor and the federal land manager are to be submitted to the President, who may approve the governor's recommendation for an SO_2 variance if the President finds it to be in the national interest.

Neither the AQRV test nor waivers and variances are applicable to Class II and III areas.

TABLE 1.4. Maximum allowable increases over the baseline for Class I areas when increments are consumed but exceeding them will not adversely affect air quality related values ($\mu g/m^3$).

Particulate matter	
Annual geometric mean	19
24-hour maximum	37
Sulfur dioxide	
Annual arithmetic mean	20
24-hour maximum	91
3-hour maximum	325

TABLE 1.5. Maximum allowable increases over the baseline in Class I areas under the variance procedure when the short-term SO_2 increments would otherwise be exceeded ($\mu g/m^3$).*

Low-terrain areas	
24-hour maximum	36
3-hour maximum	130
High-terrain areas	
24-hour maximum	62
3-hour maximum	221

A high-terrain area is one that has an elevation of 900 feet (274 m) or more above the base of the stack of the proposed facility; if such an elevation does not exist, the area is low terrain.

*Maxima to be exceeded no more than 18 days in any annual period.

Visibility Protection in Mandatory Class I Areas

Section 169A of the Act establishes as part of the PSD provisions a national goal of both preventing future impairment of visibility in mandatory Class I areas and remedying existing impairment due to man-made pollution. Visibility protection against degradation in the future is to be accomplished through the PSD provisions for review of new sources. Remedying existing impairment requires retrofitting emission control equipment to existing sources that adversely affect visibility.

Regulations to implement Section 169A (45 Fed. Reg. 80084, December 2, 1980) require the 36 states containing mandatory Class I areas where visibility has been identified as an important value to revise their SIPs in three ways: they must add provisions for using the best available retrofit technology (BART) on certain installations of existing major sources; they must adopt certain measures to supplement new source review programs regarding visibility; and they must identify and evaluate long-term strategies for achieving the national goal.

Under the visibility regulations, the state authority is required to make a determination of BART for each existing major stationary source that may reasonably be suspected of causing or contributing to visibility impairment in a mandatory Class I area. The state is required to consult with the appropriate federal land manager in identifying sources and determining BART. The determination of BART is to take into account the costs of compliance, the energy and (nonatmospheric) environmental impacts of retrofitting, existing control equipment, the remaining useful life of the facility, and the degree of improvement in visibility that may be expected to result from the application of the technology. Sources 15 years old or older at the time of passage of the 1977 amendments are exempted from the retrofit program. EPA has issued guidelines for determining BART for electric generating plants using fossil fuels and having total generating capacity greater than 750 megawatts and for other major stationary sources (U.S. EPA 1980a).

EPA has adopted a phased approach to visibility protection, with current regulations addressing impairment that can reasonably be attributed to individual sources, as in the case of plume blight. Future considerations will address problems, such as regional haze, for which drawing associations between emissions from specific sources and adverse effects is less direct.

Determining Compliance

Compliance with the PSD program is enforced through emission limitations specified in permits to construct facilities rather than through monitoring changes in ambient air quality. There are several reasons for this emphasis. First, the emission limitations are binding on the individual operators. A facility that exceeds its emission limitation is readily identified--by stack gas monitoring, for example, whereas it may be difficult if not impossible to determine the extent to which a specific source may be responsible when an increment is exceeded. Second, changes in ambient concentrations represented by the short-term increments not only challenge the state of the art of measurement but also may be masked by temporal variations in concentrations arising from natural sources or from sources in the baseline.

Moreover, actual changes in ambient air quality may not be directly related to consumption of increments. A number of restrictions in the PSD program can cause differences between the calculated status of increments and actual changes in ambient concentrations of pollutants. Certain sources--both major and nonmajor--are exempted from the PSD program. Furthermore, current regulations require that modeling computations use stack heights for facilities according to the rule of good engineering practice (44 Fed. Reg. 2608, January 12, 1979). In many circumstances, the actual heights of stacks at which pollutants are emitted to the air are greater than the GEP height, which is a function of the linear dimensions of the plant. Tall stacks help prevent ambient concentrations that would cause Class II or Class III increments to be exceeded or that would violate the NAAQS in the vicinity of large facilities by enhancing dispersion of the emitted pollutants. Dispersal, however, may exacerbate PSD problems in distant Class I areas, since the higher the point of emission, the greater the distance from the stack that pollutants are transported through the air and the greater the time for gaseous emissions to undergo physical and chemical transformations in the atmosphere.

ALTERNATIVE WAYS OF UNDERSTANDING THE PSD PROVISIONS

Based on our analyses of the PSD provisions and our study of the ways in which the program is being put into practice, there appear to be three alternative interpretations of PSD.

PSD as a System of Supplementary Standards for Attainment Areas

The baseline air quality and increments have been interpreted as establishing new national standards which, if not exceeded, will help to protect against the loss of visibility and other adverse effects on public welfare. If the provisions of the Act that are designed to promote flexibility through local decisions on classification of land, best available control technology, waivers, and variances are not used, it seems likely that the effect of the increments as ceilings will be no different than that of standards.

However, the increments were not set up as air quality standards in the usual sense of the word. Standards such as the NAAQS are absolute constraints and assume that the same concentration of an air pollutant will result in the same effects independent of location.

Under PSD, consumption of increments results in different levels of air quality in different areas because the baseline concentrations differ from place to place. Further, as described in the section on Determining Compliance, actual changes in ambient air quality are not necessarily directly related to the status of the increments.

We conclude that accepting this interpretation of PSD will be self-fulfilling: if the increments are regarded as inviolate standards, no use will be made of the features of the PSD program that provide for variations in response to local conditions and needs.

PSD as a Decision-Making Process

The PSD provisions may be interpreted as establishing the "rules of the game," i.e., a process for determining geographically variable values of air quality and thereby helping to protect against undesirable air pollution effects. Congress recognized that there is neither adequate scientific knowledge nor a consensus about the relevant social values to establish uniform standards dealing with visibility and other AQRVs. Thus, PSD defines who will have a voice in determining the relative desirability of such values, what sources will be covered, and when concerns about the degradation of these values will be triggered. Consumption of increments initiates a process for determining in individual cases the value to be placed on good air quality. PSD provides a mechanism through which each geographic area makes its own trade-offs between clean air and industrial development.

PSD as a Resource Management Tool

PSD, particularly the provisions related to BACT review, may be interpreted as expressing the judgment of Congress that clean air is a scarce resource and should be used as efficiently as possible. Economic markets do not now regulate the use of clean air. Therefore, a mechanism is necessary to evaluate whether it would be more efficient to install stringent controls on a proposed new development than to foreclose the possibility of future development by others, to help a community decide what kind of new development it wants, and to protect the public interest in areas of national importance.

On the basis of the analyses presented in this report, we have concluded that PSD should be considered a decision-making process or a resource management tool. The two interpretations emphasize different goals and different provisions of the Act but are closely related. They focus on a process rather than on a federally mandated set of absolute standards.

2
AIR QUALITY RELATED VALUES

The concept of air quality related values (AQRVs) appears in Section 165 of the PSD provisions, which charges the federal officials responsible for managing lands in Class I areas with "an affirmative responsibility to protect the air quality related values (including visibility)" of those areas.

If the responsible federal land manager demonstrates to the satisfaction of the state that a proposed facility will have an adverse effect on AQRVs on federal land in any Class I area, the facility will be denied a PSD permit even if the increments for SO_2 and TSP are not exceeded. We are not aware of any permits that have been denied because of this test. If the operator of a proposed facility applies for a waiver of the Class I increments, he must demonstrate to the satisfaction of the federal land manager that AQRVs are not adversely affected. We know of no instance of an applicant proposing such a demonstration.

The AQRV test contained in the Act applies to any federal land contained within the boundaries of any Class I area not just to mandatory Class I areas. The AQRV test does not apply to Class II and Class III areas, although lands in these areas--for example, agricultural and forest lands--could be adversely affected by degradation of air quality.

DEFINITION OF AQRVs

The term <u>air quality related values</u> is not defined explicitly either in the Clean Air Act or in subsequent

EPA PSD regulations. Because of their responsibilities under the Clean Air Act, the federal land managers (secretaries of departments containing land management agencies, such as the National Park Service, the U.S. Forest Service, and the Fish and Wildlife Service) are attempting to define AQRVs. In addition to visibility, the land managers have identified as AQRVs: odor, flora, fauna, cultural resources, geological features, climate, and acid precipitation (i.e., both soil and water qualities affected by acid precipitation) (45 Fed. Reg. 15016, April 10, 1978). Public health is generally not regarded as an AQRV because air quality in PSD areas is high and ambient concentrations of pollutants are lower than the primary NAAQS, which are intended to protect public health allowing for an adequate margin of safety.

The importance of the various AQRVs will differ from one Class I area to the next, and it is expected that the federal land managers will determine the priority of the various AQRVs on a case-by-case basis. For example, effects on flora, fauna, and ecosystems may be most important in wilderness areas, which are managed to maintain natural ecosystems, while visibility may be most important in national parks, which are managed to ensure the enjoyment of visitors. There does, however, appear to be a consensus that visibility--the only AQRV specifically mentioned in Section 165 of the Clean Air Act--is the most important AQRV. Furthermore, a part of the PSD provisions, Section 169A, deals exclusively with impairment of visibility in mandatory federal Class I areas. This subpart provides not only for protecting visibility against degradation by new sources, but also for remedying existing impairment of visibility that results from man-made air pollution.

A useful distinction can be drawn between local and regional threats to air quality. Local threats are those for which it is relatively easy to make direct associations between emissions from specific sources and effects, even though quantitative relationships may be difficult to derive. Examples of local problems include degradation of visibility by a coherent plume from a point source (plume blight) and odors emanating from gas production fields or other facilities. Associations between emissions and effects are indirect in regional problems, which result from long-range transport and physical and chemical transformation of pollutants, as well as mixing of pollutants from many sources. The most significant regional problems are large-scale haze and acid precipitation.

The PSD program, based on preconstruction analysis of the potential effects of major stationary sources, is more readily applied to local than to regional threats to AQRVs. Modeling analyses for such effects are, in principal, straightforward and deal with localized problems and at most a few sources, focusing on the behavior of plumes and disregarding emissions dispersed in mesoscale meteorological systems. In analyzing regional problems, the transport and transformation of pollutants from many sources over long distances make it difficult, if not impossible, to associate a given level of damage with emissions from a specific source. Regional air quality modeling is much more difficult than local air quality modeling. Furthermore, the entire PSD system, employing implementation plans of individual states, is inadequate for addressing multistate problems. State plans and decision making on air quality generally do not address regional problems.

The critical distinction between local and regional problems has been acknowledged in EPA's visibility regulations (45 Fed. Reg. 80084, December 2, 1980) which, because of uncertainties surrounding regional visibility problems, address only visibility problems that are "reasonably attributable" to a single source or small group of sources. EPA plans to expand the regulations in the future to include regional haze as the underlying technical understanding of the relationships between emissions and large-scale hazes improves.

TECHNICAL UNDERSTANDING OF VISIBILITY

Research on visibility over the past 50 years has resulted in a good basic technical understanding of atmospheric visibility and its relationship to ambient air quality--much better, in fact, than understanding of health effects, damage to materials, acid precipitation, or ecological effects (Charlson et al. 1978). Nonetheless, continuing research is needed, especially on existing conditions and trends in visibility and on perceptions of changes in visibility conditions.

Definition of Visibility

Visibility depends on the clarity of the atmosphere and can be defined quantitatively in terms of an atmospheric extinction coefficient (the attenuation of light per unit distance as a beam moves through the atmosphere), contrast (the relative brightness of various features in a scene), visual range (the farthest distance over which an observer is able to distinguish a large black object against the horizon sky), and discoloration (shifts in the distribution of wavelengths of light produced by pollutants in the atmosphere). These various measures of visibility, particularly the first three, are interrelated. For example, in a uniform atmosphere the visual range is inversely proportional to the extinction coefficient. The interrelationships between the various measures of visibility are important in a practical sense because an increase in atmospheric extinction affects not only the ability to distinguish distant objects but also the clarity with which one sees nearby objects. A decrease in visual range from 130 to 110 km produces a noticeable change in the contrast and coloration of a vista only 30 km away.

The choice of appropriate parameters for characterizing visibility depends on the specific problem being considered. Visual range is useful for summarizing data and for making comparisons of overall atmospheric clarity in space and time. Figure 2.1 compares median visual ranges during 1974-76 for nonurban sites throughout the United States and emphasizes the geographical differences in ambient visibility. Median yearly visual range exceeds 110 km in the Rocky Mountain West but is approximately 15 to 25 km in nonurban areas east of the Mississippi and south of the Great Lakes.

Contrast and discoloration depend on the specific features of the terrain being observed and on the distance between the observer and the terrain. Contrast and discoloration yield the best correlation with subjective human evaluations of visual air quality (Malm et al. 1979a). Thus, contrast and discoloration are appropriate parameters to consider in case-by-case evaluations of visibility effects of proposed new sources.

Atmospheric Extinction

Of all the visibility parameters, the extinction coefficient is the one that can be related most directly to the physical composition of the atmosphere. Visibility through the atmosphere is restricted by both absorption and scattering of light by both gases and

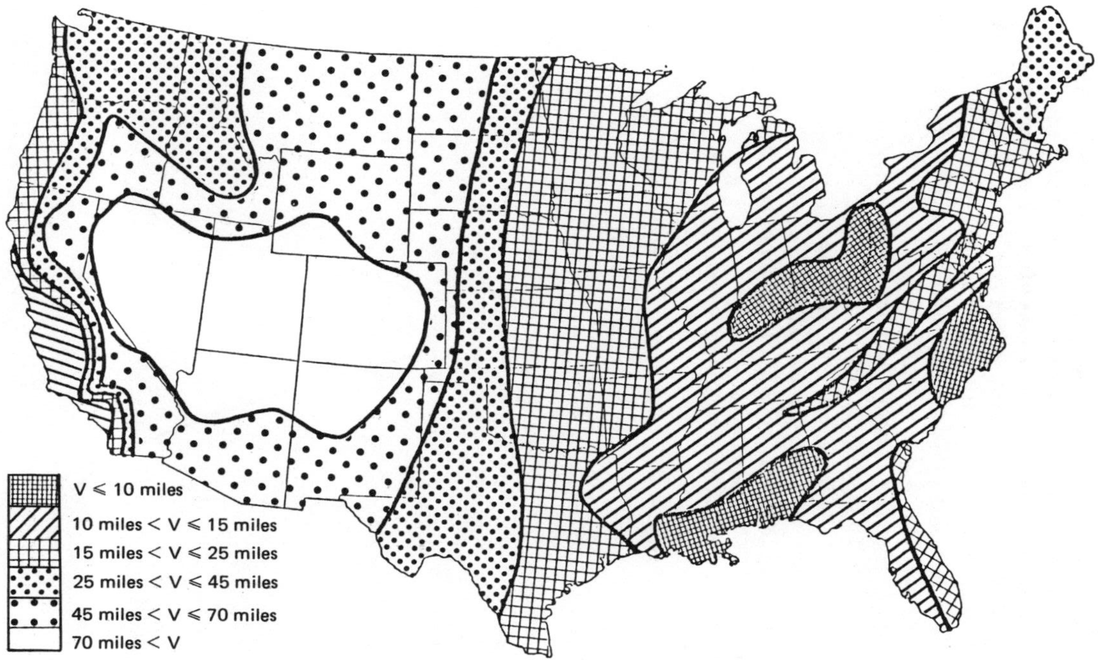

SOURCE: Trijonis and Shapland (1979).

FIGURE 2.1. Visual range in suburban and nonurban airports of the United States, 1974-1976.

particles. The extinction coefficient, then, is generally the sum of four terms.

Scattering of light by the molecules of atmospheric gases (Rayleigh or blue-sky scattering) would restrict visual range to approximately 320 to 390 km if all particles and pollutant gases were absent. Light scattering by air molecules is responsible for the blueness of the sky and for the fine blue haze observable over long distances in a clean atmosphere. Extinction due to Rayleigh scattering depends on altitude (i.e., on the density of the air).

Nitrogen dioxide (NO_2) is the only gaseous pollutant that is a significant absorber of light. Although concentrations of NO_2 are usually not large enough to reduce visual range significantly, NO_2 can produce significant brownish discoloration because it absorbs blue light preferentially.

In most cases, the dominant contribution to total extinction is from scattering of light by particles (aerosols). Figure 2.2 shows that light scattering per

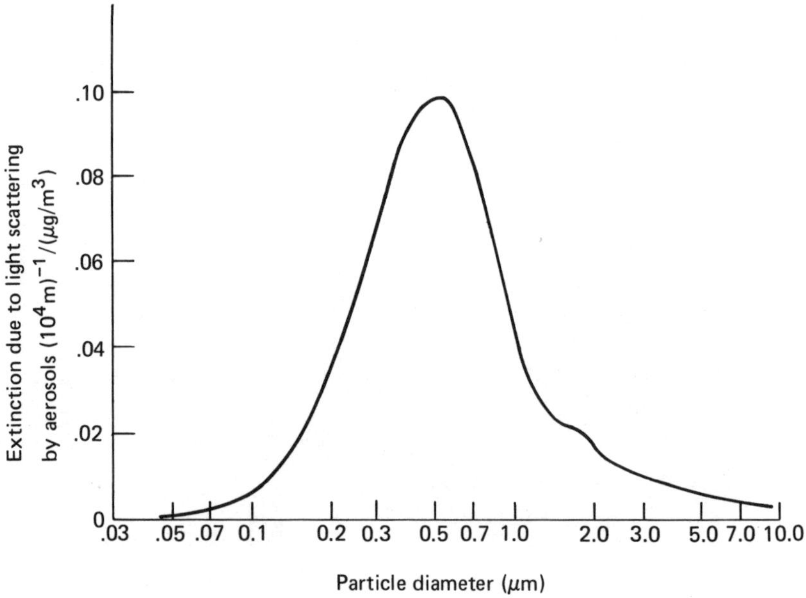

SOURCE: White and Roberts (1977) Copyright Pergamon Press, Ltd.

FIGURE 2.2. Effectiveness of light scattering by aerosols as a function of particle diameter.

unit mass of particle depends strongly on particle size; the most efficient scatterers are particles in the size range of from 0.1 to 1 micrometers. Significant contributions to light scattering can be made by particles from fuel combustion or industrial processes (such as fly ash or soot particles), aerosols from burning vegetation, fine dust particles, organic particles, and secondary aerosols (aerosols formed by gas-to-particle conversion processes in the atmosphere). Secondary aerosols, especially sulfate particles, are often the major contributors to light scattering because they tend to have diameters in the range of 0.1 to 1 micrometers. Light scattering by particles can produce discoloration as well as reductions in overall contrast or visual range; the effects of discoloration vary significantly with the relative angles of the observer and sunlight.

Particles also absorb light. The most important light-absorbing aerosol is usually graphitic carbon (soot). Absorption of light by aerosols can be substantial in plumes or urban areas where soot contributes a significant fraction of the aerosol.

As noted, light scattering by particles usually dominates total extinction in hazy air masses. The two major exceptions are remote areas of the Rocky Mountain Southwest where natural blue-sky scatter is comparable to scattering by aerosols, and certain urban areas where absorption by soot particles may be comparable to scattering by aerosols.

A basic consequence of the fundamental relationships among aerosol concentrations, extinction, contrast, and visual range is that visibility in clean areas is much more sensitive to small changes in concentrations of fine aerosols than is visibility in already polluted areas. Table 2.1 shows the effect of adding 2 $\mu g/m^3$ of sulfate aerosol--assumed to have an extinction coefficient per unit mass of .005 $km^{-1}/\mu g/m^3$--on various levels of baseline visibility. It is apparent that the relative effect is quite small when visibility is poor, but that the relative effect is significant when visibility is good.

Visibility Problems

Man-made visibility problems fall into three general categories: (1) plume blight, a plume that is

TABLE 2.1. Effect on visual range of adding 2 $\mu g/m^3$ of sulfate aerosol.

Baseline visual range [miles (km)]	Visual range with sulfate aerosol added [miles (km)]	Percentage change
200 (322)	110 (177)	45.0
120 (193)	80 (129)	33.3
80 (129)	60 (97)	25.0
50 (80)	41 (66)	18.0
25 (40)	22.7 (36.5)	9.2
15 (24)	14.1 (22.7)	6.0
10 (32)	9.6 (15.4)	4.0
5 (8.0)	4.9 (7.9)	2.0

distinguishable against the sky or terrain relatively near the source, (2) small-scale haze, local pockets of haze at ground level or in elevated layers, and (3) large-scale regional haze, widespread, regionally homogeneous haze that reduces visibility in every direction from an observer.

Modeling studies of the effects on visibility of emissions from coal-fired power plants illustrate the contributions of pollutants associated with plume blight (Latimer et al. 1978, ERT 1978). Assuming a constant wind speed of 10 km/hr, within the first 20 to 30 kilometers from the plant, visibility reduction in the plume is dominated by scattering by particulate fly ash and the discoloration effects of NO_2. Fly ash is most important if the plant lacks effective particulate controls; NO_2 is more important if particulate emissions are stringently controlled. Since most of the sulfur oxide emissions occur in the form of SO_2, a colorless gas, sulfur oxides usually do not contribute to visibility impairment near the plant. As the plume travels away from the plant, however, gaseous SO_2 becomes oxidized to form particulate sulfates. At intermediate distances and times from the source (30 to 150 kilometers, or 3 to 15 hours), sulfate aerosols and NO_2 generally determine the visible nature of the plume. At distances greater than 150 kilometers or times greater than 15 hours, sulfates become the dominant species affecting visibility.

Small-scale pockets of haze, such as in valleys and elevated layers, have not been adequately studied or characterized. However, in line with the conclusions of plume modeling studies and on the expectation that small-scale haze layers are the result of pollutants that have been in the air for some time, it is likely that secondary aerosols (sulfates and possibly nitrates) are the dominant components of these layers. In some cases direct emissions of soot, fly ash, or dust may be important to small-scale haze. NO_2 may add to the discoloration of small-scale haze layers.

Large-scale regional haze can affect areas with linear dimensions of 100 to 1,000 kilometers. Studies in several areas of the country have found that sulfate aerosol often constitutes the single most important component of the aerosol associated with regional hazes. Sulfates appear to account for approximately 40 to 65 percent of aerosol extinction in both urban and rural areas of the East (Trijonis and Yuan 1978b, Leaderer and Stolwijk 1979, Dzubay 1979, Lewis and Macias 1980, Trijonis et al. 1980) and 30 to 50 percent of aerosol extinction in nonurban areas of the desert Southwest (Trijonis 1979; Macias et al. 1979, 1980; M. Pitchford, EPA Environmental Monitoring Systems Laboratory, Las Vegas, private communication, April 1980; J. Moyers, University of Arizona, private communication, April 1980). The remainder of aerosol extinction can arise from a myriad of natural and man-made aerosols, including organic aerosols, dust, secondary nitrate aerosols, soot, fly ash, and smoke from burning vegetation. There appear to be two significant exceptions to the tendency for sulfates to be the single most important aerosol in regional hazes. One is in the Pacific Northwest where particles from burning vegetation appear to be more important (Cooper and Watson 1979, Lyons et al. 1979); the other is in certain urban areas where carbonaceous materials (both organic aerosols and soot) are more important (Courtney et al. 1980, Countess et al. 1980, Heisler et al. 1980, Wolff et al. 1980). Recent studies suggest that organic aerosols and sulfates may be comparable in their effects on visibility in the rural Southwest.

Visibility Provisions of the Clean Air Act

Section 169A of the Clean Air Act, on visibility protection for mandatory Class I federal areas, establishes a national goal of preventing future impairment and remedying existing impairment of visibility of mandatory Class I federal areas resulting from man-made pollution. In response to the requirements of Section 169A, EPA has issued visibility regulations (45 Fed. Reg. 80084, December 2, 1980) that have three basic parts: (1) application of best available retrofit technology (BART) to certain major stationary sources causing existing impairment of visibility in mandatory Class I federal areas: (2) development and continued reevaluation of long-term (10- to 15-year) strategies to remedy existing and prevent future visibility impairment in mandatory Class I federal areas; and (3) review and control of new sources to prevent adverse impacts on visibility in all Class I federal areas. The first two parts of the regulations pertain only to mandatory Class I areas; the third part, which elaborates how effects on visibility should be assessed as part of the PSD preconstruction review, pertains to all Class I areas containing federal lands.

The implementation of the visibility regulations is intended to be carried out by the individual states through their SIPs. However, the regulations guarantee a substantial role for federal land managers in consulting with states in SIP preparation, identifying vistas to be protected, defining significant impairment from existing sources and adverse impairment from future sources, and concurring on BART exemptions. Impairment of visibility is to be evaluated on a case-by-case basis for each Class I area, with consideration of both existing sources that might be subject to BART and future sources that apply for PSD permits.

In response to uncertainties about visibility problems, such as the lack of adequate models of large-scale regional haze, EPA has taken a phased approach in its visibility regulations. States are required to focus initially on clearly defined cases of visibility impairment, on evaluation and control of new sources to prevent future impairment, and on the preliminary formulation of general strategies to make reasonable progress toward the national visibility goal. In the first phase, the emphasis is on plume blight and small-scale haze. With improved scientific understanding of the relationships between emissions and impairment of visibility on a regional scale, the scope of the regulations will be extended.

Although current visibility regulations represent only the first phase of a visibility protection program, EPA has indicated in its preamble to the regulations that the basic structure of the regulations will be unchanged in future phases. We question, however, whether the current structure of the regulations will be appropriate for dealing with regional haze. Implementation of visibility regulations under the Clean Air Act is to be achieved by the individual states, through their SIPs. Controlling regional haze (which may result from interstate transport of pollutants) will require regional, if not inter-regional, coordination and institutions not envisioned in the SIP process.

TECHNICAL UNDERSTANDING OF OTHER AQRVs

Visibility is fairly well understood because many basic principles of atmospheric optics have been known for decades and because in the past few years research on the fine, optically active aerosols has been very productive. For other AQRVs, we must deal with a broader range of scientific disciplines (for example, biology, physics, and chemistry) and with water, soil, and biota as well as the air. These added complexities make it much more difficult to arrive at a fundamental understanding of many other AQRVs. We have a rudimentary understanding of other AQRVs, however, and we can expect to expand understanding considerably over the next few years.

Odor

The absence of noxious of odor is considered very important to visitors to Class I areas. It is doubtful that the criteria pollutants (such as SO_2, O_3, and NO_2) would reach concentrations high enough to cause odor problems in Class I areas. Odors are more likely to be associated with emissions of hydrogen sulfide (H_2S) (NRC 1979a), ammonia (NH_3) (NRC 1979b), and certain trace organic gases. Such emissions can arise from a wide variety of sources such as oil and gas fields, synthetic fuels plants, geothermal plants, chemical plants, petroleum refineries, paper and pulp mills, and food processing plants.

Because of the many industries and large number of specific compounds involved, it is difficult to make a general characterization of odor problems. Furthermore, the human sensory response to odorants is highly variable and sensitive. Extremely low ambient concentrations of offensive odorants can be perceived by people, but we do not yet have a scientific understanding of the physical and chemical determinants of odor and human sensory responses (NRC 1979c). Better understanding of the perception of odor is needed to help the federal land managers discharge their responsibilities under the PSD provisions.

Acid Precipitation

Deposition of acids onto plants, soils, and waters might be the most important effect of air pollution in specific Class I areas, most of which are likely to be east of the Mississippi. Acid deposition can occur in two basic forms--dry and wet. Dry deposition refers to the settling of sulfur dioxide, nitrogen dioxide, and their acidic end-products (such as sulfate aerosols, nitrate aerosols, and nitric acid gas) onto surfaces. Wet deposition refers to fall out of pollutants in rain, fog, or snow. Although both dry and wet deposition are potentially important effects, wet deposition of acids--acid precipitation--has received the greater research attention in recent years.

Acid precipitation has been linked to dramatic effects in sensitive aquatic systems (for example, to destruction of fish life in many lakes in the Adirondacks). This damage is produced not only through the direct effects of acid on fish life but also through increased concentrations of certain toxic metals (lead, aluminum, and mercury), which are dissolved by acid solutions. Acidification of lakes has ecological, human-health, aesthetic, and recreational effects (NRC 1978).

Acid precipitation can also affect soils, vegetation, and terrestial ecosystems. For soils that are deficient in sulfur and nitrogen, some deposition of these elements can be beneficial, having a fertilizing effect. Ultimately, however, acid rain can cause leaching of nutrients and accumulation of phytotoxins in soils. While there is evidence of direct adverse effects on plants, the relationship between the acidity of precipitation and crop or forest growth has not been unequivocally established. Acid precipitation may stress ecosystems as well as individual species and reduce ecological diversity; these effects could be irreversible. Much more knowledge is needed, however, concerning the effects of acid precipitation on aquatic life, soils, vegetation, and ecosystems.

Acid rain is of greatest concern in the eastern United States, because of the high levels of acidity (Figure 2.3) and because of a lack of buffering capacity in eastern soils and rock. There is also evidence that levels of acidity in the East have increased substantially since the mid-1950s (Likens et al. 1979), although this evidence is only suggestive because of uncertainties in the quality and comparability of historical data. No consistent trend is evident in data taken at nine monitoring stations in New York and Pennsylvania between 1965 and 1978 by the U.S. Geological Survey. The area of the nation where precipitation is more acidic than natural appears, however, to be growing. For example, recent measurements have shown that acid precipitation also occurs in the West

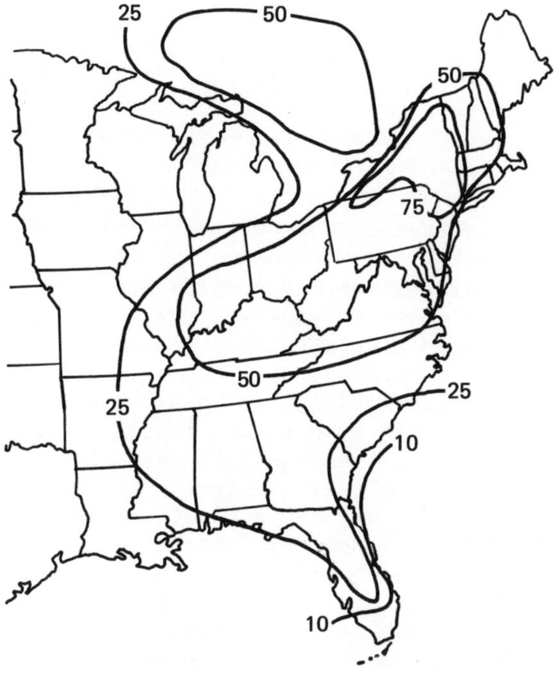

NOTE: Values higher than 2.5 microequivalents per liter (pH 5.6) are defined as acidic.

SOURCE: Likens et al. (1979).

FIGURE 2.3. Isolines of acid precipitation in the Northeast for 1975-1976 (microequivalents per liter).

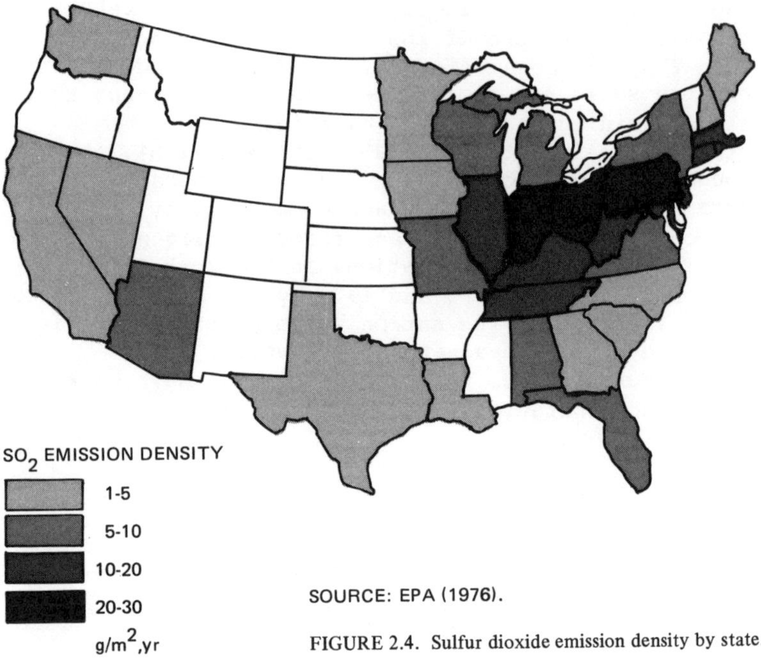

SOURCE: EPA (1976).

FIGURE 2.4. Sulfur dioxide emission density by state.

(Liljestrand and Morgan 1978, Lewis and Grant 1980) and the South (Brezonik et al. 1980).

The western acid rain problem differs from the eastern problem in several respects. Western acid precipitation is predominantly nitric acid, whereas eastern acid precipitation is approximately two-thirds sulfuric acid and one-third nitric acid. Alkaline dust particles in western air apparently neutralize much of the acidity of acid sulfates and nitrates. Because western soils are more alkaline, western lands and lakes tend to be much better buffered against the effects of acid precipitation. Furthermore, man-made emissions of sulfur oxides and nitrogen oxides per unit area are generally much lower in the West owing to the large, sparsely populated areas of that region. A thorough review of data and trends in acid precipitation is presented by GCA (1980).

Research on the complex processes by which acid precipitation is formed is expanding rapidly. Although there are complexities in the physical and chemical processes, correlations between data on emissions and data on the spatial and temporal patterns in air quality

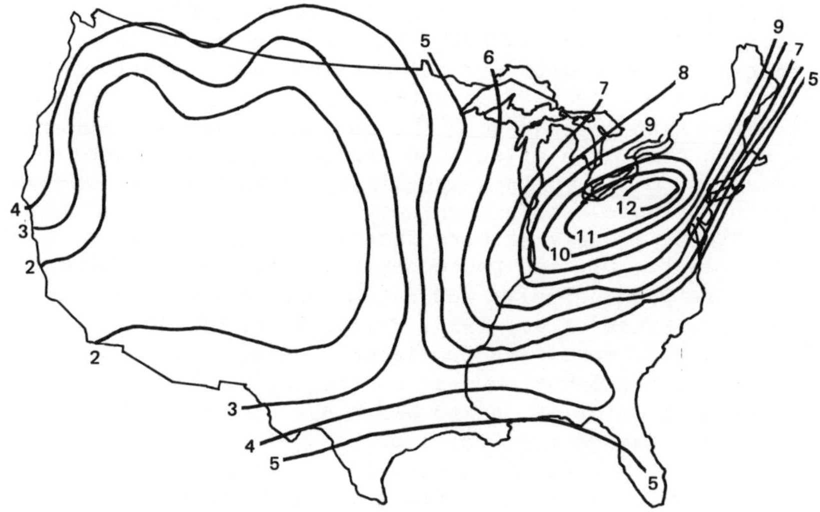

SOURCE: After Eriksson (1960).

FIGURE 2.5. Excess sulfur in precipitation over United States (kilograms per hectare per year).

strongly suggest that the fundamental causes of acid rain are man-made emissions of SO_x and NO_x. For example, correspondences among national patterns are good for: acidity of precipitation (Figure 2.3), sulfur oxide emissions (Figure 2.4), excess sulfur in precipitation (Figure 2.5), sulfate concentrations (NRC 1978, Pack 1980), nitrate concentrations (Pack, 1980), and atmospheric turbidity (Flowers et al. 1969). These patterns also show good correspondence with patterns of visual range (Figure 2.1). Natural biogenic emissions of sulfur in the eastern United States are of the order of 1 percent of anthropogenic emissions (Adams et al. 1980).

Because the acid rain problem (like regional haze and regional climate modification) appears to be large in scale, it may not be amenable to control through the SIP process. It might best be attacked by overall regional emission limitations.

Effects on Flora, Fauna, and Ecosystems

An extensive literature on the effects of air pollution on vegetation exists. Damage to vegetation produced by the criteria pollutants (such as SO_2, O_3, NO_2, or TSP) is discussed in EPA's Criteria Documents for those pollutants and elsewhere (for example NRC 1977a, 1977b, 1978, 1979d). Much of the available literature deals with effects on agronomic, forest, and other economic crops from short-term exposures to levels of pollutants greater than the secondary NAAQs. Little is known about the effects due to long-term low-level exposures to air pollutants--singly or in combination--on plants, animals, and ecosystems. However, some studies show direct adverse effects from low-level long-term exposures to SO_2, and photochemical oxidants on some sensitive plant species and native ecosystems (Brown et al. 1979, Kress et al. 1980, Miller 1980, NRC 1977b).

Injury to sensitive coniferous trees, such as pine, has frequently occurred near sources of SO_2 and has resulted in defoliation and death due to high levels of exposure. Damage from low-level long-term exposures to SO_2 is usually associated with foliar accumulations of sulfate, which may range from about 600 ppm about 2 km from a source to 100 ppm 10 km distant. At the larger distance, the more important adverse effects are decline in tree growth, changes in foliation, and alteration in mineral nutrient cycling (Legge et al. 1978). Long-term exposures to SO_2 at levels below the secondary NAAQs have resulted in changes in the composition of species and in the distribution of lichens with consequent alterations of the ecosystem (Le Blanc and Rao 1975).

Ozone damage to eastern white pine occurs after exposures of 4 to 6 hours at concentrations only one-half the secondary NAAQS. Indigenous stands of white pine have been exposed to ambient concentrations at this level for varying periods of time accumulating to as much as 185 hours over a growing period. Indigenous western ponderosa and Jeffrey pines on the western slope of the Sierra Nevada have been damaged by levels of ozone equal to the NAAQS during repeated exposures in summer months at distances up to 80 km from an urban center. Changes in foliation associated with ozone and other photochemical oxidants leads to premature needle fall, resulting in increases in pine beetle infestation, increases in ground cover and fire hazard, and decreases in seedling germination and hence stand reduction (Brown et al. 1979, Miller 1980). Susceptibility of native understorey in the western Aspen zone to low levels of ozone from nearby urban centers suggests that shifts in native vegetation may not only alter the composition of plant communities but also adversely influence watershed areas (Harward and Treshow 1975).

Comparatively little is known about the effects of air pollution on wildlife. These effects generally result from ingestion of food contaminated with fluorides, heavy metals, and other toxicants or from loss of habitat (Newman 1980a). However, a recent report demonstrates that male sweat bees exposed to SO_2 at about NAAQS levels for 16 to 29 days underwent a change in behavior expected to adversely affect mating, with consequent population changes (Ginevan et al. 1980).

To carry out their responsibility to protect flora, fauna, and ecosystems, federal land managers will need a great deal more information than is currently available about sensitive species and ecosystem processes and about susceptibility to injury or damage by exposure to low levels of pollutants. The National Park Service and U.S. Forest Service have begun work on documenting sensitive plant species, especially those sensitive to SO_2. In addition to documenting plants, animals, and ecosystems sensitive to air pollution, federal land managers should develop systems to assess the distribution and accumulation in both soils and waters of toxicants associated with emissions from fossil fuel combustion, mining, and smelting. Programs are also needed to assess the subtle effects of pollutants or combinations of pollutants on growth, senescence, productivity, and susceptability to diseases and insects.

Effects on Cultural and Geological Resources

Air pollution can have adverse effects on cultural and geological features such as archeological ruins and unusual rock formations. Damage may result from soiling, chemical interactions, and physical abrasion. There is a fair amount of literature on the effects of soiling by particulate matter and damage to building materials from acid deposition. In general, the literature, summarized in EPA's Criteria Documents, focuses on buildings, monuments, and other resources typically found in urban areas. Little is known, for example, about the effects of deposition of sulfates and nitrates on cultural and geological resources in rural areas or about physical and chemical abrasion from windblown particulate emissions (or how these effects compare with those of natural forces). For example, archeological ruins, such as those in the Chaco Canyon National Monument, are apparently deteriorating at a faster rate now than 50 years ago (W.

Herriman, National Park Service, Bloomfield, N.M., personal communication, September 10, 1980). Whether this effect is due to increased emissions in that region is unknown.

A complete physical characterization of effects on cultural and geological features is not feasible, because many pollutants may be involved, many types of materials can be affected, and effects may be directly or indirectly associated with emissions from specific sources. Assessments must be made in specific cases depending on the characteristics of the resources at risk.

Effects on Climate

Atmospheric concentrations of fine aerosols can affect climate in at least two basic ways: (1) the aerosols scatter or absorb sunlight and therefore alter the heat balance of the earth, and (2) the aerosols affect processes of cloud formation and precipitation. Although large uncertainties remain in our understanding of the effects of fine aerosols on climate, there is mounting evidence that such effects may be substantial.

It is useful to distinguish among three spatial scales of climate effects: urban, regional, and global. Urban aerosols can reduce incoming solar radiation by as much as 50 percent (McCormick and Ludwig 1967, NRC 1979d, Peterson and Flowers 1977) and can have substantial impacts on cloud formation and precipitation (Changnon 1968, Huff and Changnon 1973, Landsberg 1956, Semonin 1974). Modifications of microclimates in the vicinity of urban areas does not appear to be a concern of PSD, because the AQRV test applies to Class I areas, which are usually remote from urban areas. Furthermore, the decision in Alabama Power v. Costle has exempted sources in nonattainment areas from PSD review even if the sources affect Class I areas, except if interstate transport of pollutants is involved.

The potential of changes in global climate, due to both fine aerosols and carbon dioxide (NRC 1979d, e), is a subject of considerable public and scientific interest. Global effects are the least understood of the potential effects of air pollution on climate but, in any case, do not seem to be a direct concern of PSD.

Regional climate modification is legitimate concern in the PSD program because it can affect Class I areas. Evidence is accumulating that some regional hazes caused by sulfates and other fine aerosols, such as the haze in the eastern United States, can produce effects on solar radiation on the same order of magnitude as urban hazes--i.e., up to 50 percent reduction in solar radiation (Flowers et al. 1969, WMO 1977, Bolin and Charlson 1976, Angell and Korshover 1975, McCormick and Ludwig 1967, NRC 1979d). A reduction in the intensity of solar radiation reaching the earth's surface could have substantial effects on growing seasons and on daytime temperatures (Husar et al. 1979). Furthermore, as in urban areas, there may be regional effects on cloud formation and precipitation.

ECONOMIC VALUATION OF AQRVs

The problem of estimating the benefits of protecting AQRVs is being treated in an expanding economics literature. This section summarizes some recent findings on visibility and acid rain and presents some major conclusions of economic benefit assessments.

The Value of Atmospheric Visibility

A number of empirical studies of the value of atmospheric visibility are available (Brookshire et al. 1976, Loehman et al. 1980, Randall et al. 1974, Rowe et al. 1980, Schulze et al. 1980). Rather than using observed market behavior to acquire the data for analysis, these studies have used bidding games, an untraditional economic technique originally developed by Davis (1963). The technique, based on the economic theory of consumer behavior, uses structured interviews to establish directly a measure of welfare change--the willingness to pay or to be compensated for a change in environmental quality (Bradford 1970). Because of differences in the measures of value used, locations studied, and representations of atmospheric visibility, the studies are not fully comparable.

Table 2.2 contains results of two studies that are reasonably comparable. Each study was performed in the Four Corners area of the Southwest; each attempted to value visual range between approximately 40 and 80 km, and each asked 100 or more current residents and visitors to the area what they were willing to pay for improvements in air quality. The two studies differed in

TABLE 2.2. Comparison of studies valuing visibility in the Southwest.

Study	Goods being valued	Location	Bidding vehicle	Mean annual household willingness to pay for a change from– High-to-medium damages	Medium-to-low damages
Randall et al. (1974)	Visibility, spoil banks, and transmission lines	Farmington, New Mexico	Sales tax	\$35	\$50
Rowe et al. (1980)	Visibility	Farmington, New Mexico	Utility bills	\$25	\$57

the form of payment hypothesized, and Randall and co-workers depicted spoil banks and transmission lines (that is, changed objects in the scene as well as visual range) in the vistas interviewees were asked to value. Although Schulze and co-workers (1980) collected evidence that the values individuals in other parts of the country attach to preserving visibility in some national parks of the Southwest far exceed the values users attach at the times of their visits, neither of the studies in Table 2.2 considered the valuations of individuals not in the Four Corners region at the time of the study.

The extent to which the figures in Table 2.2 represent the values the people of the United States attach to visibility in the Southwest is arguable; preference experiments are controversial. The numbers in the table may not be reliable indicators of absolute value. Table 2.2 does, however, indicate the nature of the relative changes in values attached to changes in visibility (measured as visual range). The value of improvements in visibility apparently depends on prevalent conditions of visibility. In particular, if visibility is poor, a bit of improvement does not matter as much as when visibility is already fairly good. Schulze et al. (1980) found a similar but more pronounced pattern. The same pattern may occur in valuations of lushness of forest vegetation (Buhyoff 1978). In other words, willingness to pay for a given increase in visibility declines with deteriorations in visibility. If further work confirms the hypothesis, compromise solutions--in which intermediate levels of pollution are permitted in an attempt to balance the benefits of preventing deterioration of visibility and the costs of controlling the responsible emissions--may not make sense economically. Economics may justify only stringent controls where existing visibility is very good or minimal or no controls where visibility is already impaired.

The Value of Controlling Acid Precipitation

On the basis of a synthesis of relevant literature in the physical and biological sciences and in stock, price, and output information, Crocker et al. (1980) estimate that the value of the effects of acid precipitation on the annual yields of existing economic activities in the eastern third of the United States in 1978 was approximately $5 billion. They estimate that $2 billion were lost in effects on materials, $1.75 billion in damages to forest ecosystems, $1 billion in direct effects on agriculture, $0.25 billion in effects on aquatic ecosystems, and $0.10 billion in other effects, including on health and water supply systems. For the most part, the estimates are incomplete because they disregard changes in prices, activities, and locations induced by acid precipitation.

As precipitation continues to be acidic, the relative ranking of economic effects could readily change. For example, increases in the geographical range and severity of acidification of lakes and streams could result in substantial losses to recreational and commercial fishing. Because of the water and soil treatment facilities already in place that can readily be adapted to handle liming procedures, the costs of acid precipitation on human health, agricultural soils, and household, commercial, and industrial water supply systems are now and are likely to continue to be insignificant compared to the other effects. The usefulness of liming of either forests or bodies of water on a large scale awaits determination.

However, this simple accounting of the costs of acid precipitation neglects at least two important features of the effects of acid precipitation on ecosystems.

First, empirical evidence suggests that unacidified aquatic ecosystems are often affected much more dramatically by a given increase in acidity than are systems that are already acidified (Raddum 1978, Butler et al. 1973). Butler and co-workers (1973) studied Pennsylvania streams affected by acid mine drainage; they found that the effects of increasing levels of acidity both on sections of streams containing fish and on the number of species present depended strongly on the acidity of the receiving waters. In other words, if the system is clean, a bit of acid does a great deal of harm; but if it is already acidic, a bit more or a bit less acid will not matter much.

These findings, and others like them for aquatic ecosystems, imply that incremental benefits increase with progressively more stringent controls on acid precursors. Where an environment is already acidified, to reduce its state of acidity could be more expensive and provide fewer interim benefits than protecting a nonacidified state. As in the case of visibility, decisions to control acid precipitation may have a strong "all-or-nothing" character: intermediate control measures can lead to burdensome control costs while generating few environmental benefits.

A second feature of acid precipitation not treated by Crocker et al. (1980) follows from the fact that effects on terrestrial and aquatic ecosystems may be irreversible. Leaching of plant nutrients, concentration of phytotoxins (including heavy metals), destruction of nitrogen-fixing bacteria, and spoliation of soil-binding agents are irreversible effects induced by acidification. With continued exposure, the built-in buffers of natural ecosystems are irreversibly (permanently) being worn away.

While the effects of air pollution on visibility are reversible, an impact of acid precipitation is the depletion of future economic opportunities. Assessments that consider only reductions in current yields from current economic activities and neglect the loss of opportunities in the future disregard an important economic feature of the effects of acid precipitation.

Irreversible effects on ecosystems and incremental benefits that grow as acid precipitation or visibility degradation is curtailed suggest that using only the economic values of losses to current economic activities can underestimate the benefits of pollution control. To address the problems of acid rain and degradation of visibility comprehensively, decision makers may have to consider large changes as well as small ones and losses of future opportunities to the material and life-supporting services provided by ecosystems in addition to current costs.

MONITORING AQRVs

Neither the Clean Air Act nor federal PSD regulations include specific requirements for monitoring AQRVs. The federal PSD program contains provisions for monitoring, but only of the pollutants regulated under the Act. There are no regulatory requirements for monitoring either the noncriteria pollutants that are most directly related to AQRVs or the AQRVs themselves. In response to the concern over AQRVs, however, monitoring programs have been initiated by EPA, federal land managers, and the industrial sector. This section provides a brief overview of monitoring data for AQRVs.

Monitoring Visibility

There are five basic ways of measuring visibility: human observation, telephotometry, photographic photometry, transmissometry, and nephelometry. Observation provides a direct but crude measurement of visual range, which can be converted to an estimate of extinction coefficient using the simple reciprocal relationship between visual range and extinction. Human observation of prevailing visibility entails viewing a series of markers (for example, buildings, mountains, or towers) against the horizon sky. A telephotometer, a device consisting essentially of a photometer combined with a telescope, can be used to measure the contrast between a target and the background sky. If the inherent contrast (contrast at zero distance) and the distance to the target are known, simple formulas allow visual range and extinction to be calculated from the contrast measurement. Photographic photometry is very similar to telephotometry except that contrast changes are measured on photographic film using a densitometer. A transmissometer measures the transmittance of light over a given path length, and thus provides a direct measurement of the total extinction coefficient. Nephelometers provide a direct measurement of one component of the extinction coefficient, the scattering component. Discussions of the advantages and disadvantages of the various methods of measuring visibility can be found in Tombach (1978, 1979), EPA (1979a), and Malm et al. (1979b).

The largest data base available for examining visibility consists of human observations of "prevailing visibility" made hourly at hundreds of weather stations (mostly airports) throughout the country over the past few decades. Because of the rather crude nature of the measuring technique, and because of inadequacies in available markers and inconsistencies in reporting practices, these data should be interpreted with caution (Trijonis 1979, 1980). Despite the limitations, several studies have found that airport data are of adequate quality for establishing relationships between visibility and meteorology, overall geographical patterns of visibility, seasonal patterns of visibility, relationships between visibility and concentrations of aerosols, and historical trends in haziness (Trijonis 1979, 1980; Trijonis and Yuan 1978a,b; Trijonis and Shapland 1979; Husar et al. 1976; Husar et al. 1979; Latimer et al. 1978; Holzworth 1962; Keith 1964, 1970; Leaderer and Stolwijk 1979).

To characterize visibility in Class I areas accurately, more precise measurements are needed than those available from weather stations. Monitoring for visibility should include color photographs, human observations, and telephotometry data (U.S. EPA 1979a). Where possible, these should be supplemented with nephelometry data, size-segregated particulate data, NO_2 measurements, and weather data. In the past few years this more detailed type of monitoring has begun. Networks are operated by EPA and the National Park Service in the West (using telephotometers, nephelometers, photographs, particulate samplers, and weather data), by the Electric Power Research Institute both in the East and in the West (using telephotometers, nephelometers, cameras, human observations, particulate samplers, and weather recorders), and by contractors at sites with potential for industrial development. In a few years these networks will be able to provide much more detailed information on visibility than has heretofore been available. It will be important to conduct extensive analyses comparing these data with one another and with the airport data. To date, a major shortcoming of many air pollution measurement programs, including visibility studies, has been the tendency to devote too few resources to data analysis as compared to data collection.

Monitoring Acid Precipitation

Historically, the monitoring of acid precipitation in the United States has been rather sparse (Likens et al. 1979); data have been available only from several limited and disjointed programs. Because of the growing concern over acid precipitation, much more data are now being collected. Examples are the federal MAP3S project, individual state programs, Canadian programs, and industry studies (U.S. EPA 1979b). A number of monitoring programs are coordinated through the National Acid Deposition Program, a voluntary association of researchers in the field. The data from these studies, when adequately analyzed, should provide a more thorough characterization of the existing acid rain problem.

Measuring the acidity of precipitation using a network of samplers presents a number of challenges. Means must be found for protecting equipment in the field from contamination and for minimizing losses from samples before collection and analysis (Hales and Dana 1979). In addition to quality control of measurement, networks need adequate regional coverage, more extensive analysis of chemical species, and improved meteorological and climatological data (U.S. EPA 1979b).

Monitoring Other AQRVs

Monitoring ambient air quality is useful in ascribing short-term direct adverse effects of selected pollutants on AQRVs (such as injury to vegetation and local climate modification), but such monitoring is of little value in assessing long-term chronic effects on animals, plants, and ecosystems; possible long-scale climate change; and alteration of geologic and historic cultural features. Detection, measurement, and evaluation of chronic and long-term adverse effects of airborne pollutants on AQRVs in biological and physical systems is best done through long-term surveillance of animals, plants, ecosystems, and materials. Ecological effects can be monitored by using indicator species that show shifts in composition and distribution, alterations in growth and productivity, or accumulation of residuals (NRC 1977c). Monitoring for AQRVs associated with biological systems calls for detailed studies of baseline conditions aimed at describing the dynamics of plant communities, their structure, composition (Treshow and Allan 1979), and associated fauna.

AQRVs AND THE PSD INCREMENT SYSTEM

In establishing the current PSD system, Congress defined specific increments of degradation for SO_2 and TSP and required EPA to formulate PSD regulations that would apply to other criteria pollutants--that is, hydrocarbons, carbon monoxide, ozone, lead, and nitrogen dioxide. Both the averaging times and numerical values of the SO_2 and TSP increments appear to have been chosen without reference to specific relationships with AQRVs. The averaging times in the current program were apparently chosen merely because they are the averaging times for NAAQS for the respective pollutants. It is, for example, unclear just what the consequences for AQRVs would be if the 3-hour increments for SO_2 and TSP were exceeded more than once a year--particularly since baseline concentrations vary from place to place. Because emissions of SO_2 and TSP bear only indirect relationships to AQRVs, we conclude that the SO_2 and TSP increments are not substitutes for the effects-related test provided by the Act (see Chapter 1 on the AQRV test for Class I areas) and that the increments should thus be seen only as signals in the PSD process.

EPA is currently developing PSD regulations for the Set II pollutants. There is a serious question, however, whether or not the AQRVs of interest are directly affected by the specific criteria pollutants. The AQRVs in Class I areas are often more directly affected by non-criteria pollutants or synergistic combinations of criteria pollutants.

Visibility

The current set of PSD increments for SO_2 and TSP have little direct association with visibility. Sulfur dioxide is an invisible gas. Total suspended particulate matter does not bear a close relationship with visibility because the size and number of the particles, not their mass, are the critical factors governing optical effects. For example, if we assumed that the 24-hour TSP increment, 10 $\mu g/m^3$, were composed only of coarse particles--greater than 2.5 micrometers (μm) in diameter--then adding this amount to relatively clean air would produce a moderate visibility effect: a background visibility of about 130 km would be reduced to approximately 108 km. If, however, the particulate matter consisted of aerosols in the optically critical size range--0.1 to 1.0 μm--then adding 10 $\mu g/m^3$ would produce an extreme visibility effect: a background visibility of about 130 km would be reduced to approximately 44 km.

Visibility, then, tends to be controlled by concentrations of fine particulate matter, and the size and chemical composition of the particulate matter are extremely important. Thus, establishment of a PSD increment for fine particle concentrations (diameters less than about 2.5 μm) would be the best way to make preconstruction reviews more relevant to visibility. Increments for fine particles themselves could take account of direct emissions of particles, such as graphitic soot, as well as particles formed in the air from gaseous precursors, such as sulfate aerosols.

Most of the Set II pollutants bear no direct relation to visibility. Carbon monoxide, ozone, and gaseous hydrocarbons are all invisible gases, but the presence of ozone affects transformation rates of SO_2 to sulfates. The only Set II pollutant directly associated with visibility is NO_2, a brownish-colored gas. Nitrogen dioxide is a significant contributor to the brown coloration of plume blight within the first few tens of kilometers from major sources. Nitrogen dioxide might also contribute to some of the observed brownish color of small-scale hazes and urban hazes, although much of the brownish color of hazes is due to preferential scattering of red light by aerosols.

Therefore, another step that could be taken to improve the effectiveness of preconstruction reviews for protecting visibility from degradation due to plume blight would be to establish an increment for NO_2. However, because major new sources are likely to be located more than a few tens of kilometers from Class I areas, an NO_2 increment is not as important for visibility in Class I areas as a fine particle increment.

Other AQRVs

Sulfur dioxide and NO_2 contribute to acid precipitation through their transformation products, sulfate aerosols, nitrate aerosols, and nitric acid vapor. Climate modification is also more closely tied to fine particles than to TSP.

In some cases, SO_2, TSP, and the Set II pollutants do bear direct relationships to other AQRVs. Certain well-documented effects on flora have been associated with particulate matter, NO_2, SO_2, and ozone at concentrations above the secondary NAAQS. Many effects on flora and fauna, however, may be more closely related to low level long-term exposures and to atmospheric transformation products of SO_2 and NO_x rather than to their precursor gases, the criteria pollutants. High concentrations of SO_2, NO_2, and ozone can also produce odors. But at the low concentrations expected in Class I areas, only H_2S and specific organic gases are expected to produce detectable odors (Leonardos et al. 1969, NRC 1979a).

PROTECTION OF AQRVs BY THE CURRENT PSD REGULATIONS

Although the relationships between the PSD increments and degradation of AQRVs is only indirect, the current PSD program does provide some direct protection of visibility and other AQRVs. The AQRV test for new or modified major sources affecting Class I areas requires direct consideration of AQRVs, and Section 169A deals directly with visibility.

Several critical weaknesses exist in the direct protection mechanisms. First, because of inadequate models for visibility and other AQRVs and because of the scientific uncertainties concerning some of the AQRVs, it may not be possible for the federal land managers or applicants to apply the AQRV test effectively. Second, Phase I of the EPA visibility regulations does not address the problems of deterioration of visibility from large-scale regional haze. Moreover, it is not clear that future phases of the visibility regulations can deal adequately with regional haze. Third, it seems doubtful that the large-scale problem of acid precipitation can be addressed by considerations of effects at individual Class I areas.

The present PSD system also provides some indirect protection for AQRVs. The most important example is the requirement that new sources use the best available control technology (see Chapter 5). Application of BACT helps to ensure that emission rates from new sources are minimized, within constraints of considerations of energy consumption, economic impacts, and environmental effects. To the extent that emissions of gaseous precursors of sulfate and nitrate aerosols are controlled, protection is provided against degradation of visibility, acid precipitation, and deterioration of other AQRVs.

The parts of the PSD program which require stringent emission controls protect AQRVs from new sources of emissions only. The main contributors to SO_2 emissions in the East in the 1990s will be plants that were already in existence by 1976 (MITRE 1979). Thus, at least for the next decade, stringent emission controls for new sources will not solve the problems of regional haze and acid precipitation; they will only keep emissions that contribute to those problems from getting worse. Additional direct protection for regional haze and acid precipitation could be achieved by regional emission limitations for SO_2 and NO_x and by encouraging the replacement of older, less well controlled plants with new facilities.

3
AIR QUALITY MODELING

Under Section 165(a)(3) of the 1977 amendments to the Clean Air Act, applicants for PSD permits must demonstrate that their proposed emissions will neither cause nor contribute to ambient concentrations exceeding either the allowable PSD increments or the NAAQS. The PSD requirements also limit the amount by which ground-level concentrations of SO_2 and TSP may increase over baseline concentrations as a result of emissions from most new sources, whether or not they are subject to preconstruction review as major sources.

Air quality models are the only tools available for relating expected emission rates of pollutants to their effects on ground-level concentrations. Estimating the potential degradation of AQRVs, such as visibility, also requires appropriate models, since the first step in such an estimate is to calculate ambient concentrations. In addition, modeling is needed to keep account of the status of the PSD increments since, as explained in Chapter 1, monitoring generally cannot be used for that purpose.

Several types of models, performing different functions, are needed to carry out the analyses inherent in the PSD program. Point-source models are used to determine the effect of emissions from a single source on ambient concentrations of pollutants. Point-source models include general screening models, needed to determine whether proposed sources warrant more detailed--and more expensive--analysis, and more specific and highly refined models, needed to assess more accurately the impacts of those sources identified by the screening procedure as being of concern. Regional models, incorporating a considerable number of point-source emissions as well as emissions from areawide or distributed sources are needed to keep track of the status of PSD increments. Finally, models are needed to estimate the consequences of changes in ambient concentrations on AQRVs, including the human perception of these consequences.

ISSUES

It is both economically and administratively desirable to have uniform procedures for using a relatively small number of models to estimate impacts on air quality. In an attempt to achieve nationwide consistency in atmospheric modeling, the 1977 amendments directed EPA to promulgate regulations specifying with reasonable particularity the air quality model or models to be used under various sets of conditions in PSD analyses [Section 165(e)(3)(D), 42 USC 7475]. EPA complied by issuing a "Guideline on Air Quality Models" (U.S. EPA 1978), which was incorporated in its PSD regulations. Applicants are required to use the models recommended in it. Recently, EPA proposed revisions to its modeling guidelines; among other changes, the proposed revisions would expand the number of approved models (U.S. EPA 1980c). Applicants seeking to use a model other than the recommended models may do so only upon a finding by EPA that the proposed model is "comparable" to the guideline model.

Three basic policy issues surrounded implementation of modeling guidelines: (1) the need for screening protocols to determine whether full-scale PSD modeling analysis is warranted in each case; (2) the need for a cutoff point, based upon distance from the source or upon air quality impact, beyond which model estimates would not be needed; and (3) the need to create an arbitration board to resolve PSD modeling disputes[1] (Raffle 1979).

---

[1]Industry petitioners in Alabama Power v. Costle also claimed that EPA's modeling guidelines inhibit the adoption of newly proposed, potentially more accurate models and that the modeling guidelines were unnecessarily conservative. The court ruled, however, that EPA met its obligations under the Federal Administrative Procedures Act in promulgating its

A screening technique involves the application of a conservative air quality model to identify, with minimum effort, those sources that clearly will not cause the increments or NAAQS to be exceeded. Refined modeling analyses are then required only when screening techniques indicate that increases in ambient pollutant levels would consume a pre-selected fraction of the remaining increment or violate an ambient standard.

The second policy issue is related to the ability of available models to estimate low-level, long-range air quality impacts accurately. Dispersion characteristics commonly used in available models are based on empirical data derived from short-range experiments. Attempts to extrapolate short-range relationships to greater distances leave the results open to technical challenge. Consequently, EPA decided to limit the general application of the air quality models contained in the original guideline document to downwind distance of no more than 50 km (31 miles), but the limit should not be construed as implying that sources do not affect air quality at distances greater than 50 km. The proposed revisions in the modeling guidelines discuss models for long-range transport but do not propose a specific model.

The third issue arises because industry representatives want to have a mechanism for resolving disputes in cases where refined modeling techniques are not readily available and where professional judgment must play a significant role. EPA has agreed to establish an external advisory group to resolve recurring difficulties regarding consistent regional or state interpretation and application of the guidelines (Raffle 1979).

CURRENT PRACTICE

Gaussian Models

The EPA guideline models are all based on the Gaussian formula for the steady-state, three-dimensional distribution of a pollutant emitted from a point source.

---

modeling regulations. Nevertheless, the ruling called on EPA "to proceed in the (modeling) area with care and flexibility appropriate to such intricate and novel subject matter."

The Gaussian formulation is derived from the general principle of conservation of mass of an atmospheric species. The major advantage of Gaussian models is their relative simplicity; little computational effort and input data are required to produce an estimate of the ground-level concentration of a species downwind of a proposed source.

The guideline Gaussian models are all steady-state models. Over the time interval considered in the calculation, factors such as wind speed and direction, temperature, emission rate, and mixing height (the height above which a spreading plume does not penetrate) are assumed to be constant. Critical parameters in Gaussian models are the horizontal and vertical dispersion coefficients, σ_y and σ_z respectively, which are measures of how the plume spreads as a function of distance from the source. The usual approaches use empirical relationships for the dependence of the dispersion coefficients on downwind distance and the stability of the atmosphere.

As empirical parameters, the dispersion coefficients must be determined experimentally. Two sets of values for the coefficients, the Pasquill-Gifford (PG) and the American Society for Mechanical Engineers (ASME) coefficients, are in general use (Hanna et al. 1977). The PG coefficients, which were originally determined for 3- to 10-minute averaging times for dispersion of a tracer released at ground level over flat terrain, are to be used in the EPA Gaussian models as hourly dispersion coefficients. Although the dispersion coefficients were determined from experiments over flat terrain, Gaussian models have been applied in cases where the terrain is complex. (See the next section, on Modeling in Complex Terrain.)

Improved data can be incorporated into the Gaussian dispersion formulation. For example, measurements of the statistical fluctuation in wind direction can provide the basis for a better estimate of the horizontal dispersion coefficient, σ_y, than stability class. Nevertheless, certain assumptions inherent in the Gaussian formula, such as steady state conditions and an average wind velocity that is not a function of altitude, cannot be improved in the Gaussian framework.

Gaussian approaches are totally inapplicable to problems involving nonlinear chemical reactions, such as those that determine photochemical oxidant levels resulting from emissions of hydrocarbons and nitrogen oxides. Because the transformation of SO_2 to sulfate is frequently represented as a first-order chemical reaction, Gaussian models can be used to model both SO_2 and sulfate. Although the conversion of NO to NO_2 and other pollutants containing nitrogen is a complex nonlinear process, the Gaussian approximation can be used in isolated plumes where NO is oxidized to NO_2 by ambient ozone.

Alternatives to Gaussian models are generally also based on conservation of mass but with fewer restrictive assumptions. Whereas the Gaussian formulation is the result of assumptions that make possible an analytical solution of the basic conservation equation, the more detailed models generally require a numerical solution of the equation, integrating the equation step by step over time. Unlike the Gaussian model, more complex models trace the movement of a plume through time and space, following changes in wind speed, direction, and atmospheric stability. Gaussian models, in contrast, use an overall description of wind and stability to produce a statistical, steady-state picture of the plume as a function of distance from the source. Non-Gaussian models are based on numerical solutions of the turbulent diffusion coefficients. These coefficients vary in a complicated manner with height and meteorological parameters. The numerical solutions may lead to dubious results, such as infinite or negative coefficients.

Modeling in Complex Terrain

Modeling dispersion of pollutants in complex terrain presents problems of describing the behavior of coherent plumes that encounter substantial changes in elevation of terrain some distance from the point of emission. In its revised guidelines (U.S. EPA 1980c), EPA recommends the VALLEY model as a screening technique in complex terrain to determine annual average and 24-hour maximum concentrations.

In the VALLEY model (Bass et al. 1978), plume trajectories are treated differently under stable and unstable conditions. For stable atmospheric conditions, VALLEY algorithms assume the plume trajectory to be horizontal until the plume comes within 10 m of the terrain. The plume is then assumed to follow the terrain at a height of 10 m. In practice, the 10-m stand-off distance does not reduce ground-level concentrations by more than 2 to 3 percent below those assumed to result from direct impact. VALLEY also assumes a linear decrease of concentration to zero at elevations approximately 400 m above the elevation of impact. Because only maximum values of ground-level concentrations are used for regulatory purposes, the linear decrease with elevation above the plume is of no consequence to regulatory decisions. VALLEY predicts the largest ground level concentrations under stable conditions--generally the constraining conditions for air quality permitting decisions in rough terrain.

For neutral and unstable conditions the VALLEY model assumes that the plume rises over the terrain to a local elevation equal to the height of the plume above the stack base at the source. The assumption leads to the same estimates of maximum ground-level concentrations over complex terrain as over flat terrain. In many neutral-flow circumstances, estimates such as these are unrealistically low, because under neutral conditions the plume trajectory does not follow the terrain completely. The PG coefficients are used in the VALLEY model only for vertical diffusion; it is assumed that the plume spreads horizontally in a 22.5° sector (centered on one of 16 wind directions). For conservatism under these atmospheric conditions, six hours of persistent 2.5 m/s winds and stable atmospheric conditions are assumed. This assumption, however, can be relaxed by using actual meteorological data.

Although the VALLEY model is recommended because no other complex terrain model is available, the steady-state Gaussian formula is inappropriate for representing flow in areas of rough terrain. Furthermore, the basic assumption that the flow field in rough terrain can be characterized as either following the terrain or directly impinging on it is an oversimplification.

Because of the high costs and long time required for a full-scale investigation of the suitability of a new or expanded site, a screening model suitable for complex terrain--simple enough to give a rapid initial identification of potentially unacceptable effects, yet accurate enough to avoid eliminating sites that would be found acceptable if analyzed in more detail--is highly desirable. Without more refined models, there is a danger that using only a simple screening model, such as VALLEY, can cause a site to be rejected without consideration of the actual meteorology, topography, or other site-specific characteristics that influence plume transport, dispersion, and impact on elevated terrain.

Moreover, the VALLEY model can generate substantial differences in results, depending on the stability class used. The great sensitivity of the model to the data fed into it may lead to arbitrary decision making. Given the large uncertainty in the VALLEY model predictions for certain elevated terrain situations, the model is open to legitimate technical challenge, which can lead to delays in the site-selection process.

The application of complex-terrain diffusion models has received considerable attention from the scientific community. At the Specialists Conference on the EPA Modeling Guideline held at the Argonne National Laboratory, February 22 to 24, 1977, there was unanimous agreement that no general complex terrain model could be recommended (Bass et al. 1978). The group was able to reach consensus on three issues:

1. High priority must be given to developing validated models so that thorough efforts to evaluate effects of sources in complex terrain can be made.
2. A specific administrative procedure must be adopted, including a two-level hierarchy of models for use in complex terrain, consisting of both conservative screening procedures and, when indicated, detailed analytical studies leading to a "rational and adequate assessment."
3. Specific screening procedures must be developed for different physical situations, including cases of the alteration of the plume trajectory when the plume is higher than terrain, alteration of diffusion rates in the near-fields as opposed to the far-fields of sources, and special flow zones such as in channeled terrain or on the lee side of hills or bluffs.

At the American Meteorological Society Workshop on Stability Classification Schemes and Sigma Curves (Boston, June 27 to 29, 1977) it was recommended that estimates for the dispersion coefficients be modified to account for both distortion of the plume and enhanced dilution from turbulence generated by the contours of the terrain (Bass et al. 1978).

Long-Range Transport Models

Predicting increment consumption in Class I areas often involves transport beyond the recommended maximum distance of application of the original set of guideline models (50 km). The fact is that long-range transport and large-scale dispersion have important consequences for PSD goals. Visibility degradation from regional haze is the eventual result of long-range transport and transformation of sulfur pollutants, and acid precipitation results from long-range transport and transformation of sulfur, nitrogen, and other pollutants. Long-range transport models provide data for analyzing degradation of AQRVs in many PSD cases.

Transport

Virtually all models of atmospheric dispersion and transport use available meteorological data in constructing a flow field to represent the motion of the atmosphere. Pollutants are then treated as being imbedded in the constructed flow field. The motion of the air itself is not calculated from first principles; only the behavior of the pollutants is tracked.

Different types of models are useful for the description of transport and dispersion over various horizontal scales. For distances on the order of 10 km, Gaussian plume models are usually appropriate under steady atmospheric conditions over flat terrain. On horizontal scales of the order of 100 to 1,000 km, however, the Gaussian formulation is not a valid description of transport and diffusion. For such large scales, two general types are available (Eliassen 1980):

- models giving, in principle, concentrations on a daily or more frequent basis by calculating solutions to a deterministic equation, and
- models of a statistical nature, giving only average concentrations for longer periods, e.g., seasons or years.

Models that calculate short-term concentrations are generally based on the unsteady state atmospheric diffusion equation. Many of the models in this category that have so far been applied are one-layer models describing horizontal transport and dispersion of pollutants in a boundary layer of constant height that is well mixed in the vertical direction (Eliassen 1980). The assumption of a well-mixed vertical distribution of pollutant concentrations is generally valid for distances of the order of 100 km or more. (For transport over less than 100 km, pollutants will not have had time to mix completely throughout the boundary layer, and in this case models should include a description of vertical concentration distributions.) The assumption of full vertical mixing is not valid in the case of transport of elevated stable layers, which is particularly important in large-scale pollution episodes.

One form of the dynamic model for long-range transport follows the transport of a small burst or puff of emissions. These models of moving parcels of air assume the puffs undergo Gaussian diffusion as they are transported in the flow field. MESOPUFF, contained in the proposed revisions to the guidelines, is one such model.

Many of the models giving calculated pollutant concentrations on a daily or more frequent basis can be used to generate average concentrations for longer periods by averaging the daily results. The second general class of long-range transport models, however, is designed to provide long-period averages only. This class is often based on a steady state form of the atmospheric diffusion equation averaged over the meteorological conditions for the period of interest, usually a year.

Atmospheric Processes

In addition to transport and diffusion, the atmospheric processes of concern to long-range models are removal and transformation. Removal of a pollutant from the atmosphere can occur either by dry deposition (i.e., absorption at a surface) or by wet deposition (i.e., absorption and precipitation by rain, fog, or snow). In addition, species can be transformed by chemical reactions with other airborne species. The EPA guideline models generally do not include provisions for dry and wet removal or for transformation. Over transport distances on the order of 10 km, little removal is expected to occur, and the omission of removal processes from models does not introduce serious errors into computed concentrations. For distances of 100 to 1,000 km, however, the residence time of pollutants is sufficiently long that removal processes can have a substantial effect on airborne concentrations.

All long-range transport models in current use include descriptions of removal processes. In one-layer models that have uniform vertical concentration distributions, the rate of removal by dry deposition is usually assumed to be proportional to the concentration. The proportionality factor is different for different species and also depends on the level of turbulent mixing and the state of the receiving surface. Wet deposition rates are usually taken to be either constant during precipitation or proportional to the precipitation rate. Estimated values for wet deposition rates are more uncertain than those for dry deposition because of lack of precipitation data and difficulties in describing rain scavenging.

When transport of sulfur compounds is considered, models should include chemical transformation of SO_2 to sulfate. The common procedure is to assume transformation proportional to the SO_2 concentration; the transformation rate depends on which of the possible chemical pathways dominate the process. Transformation rates used in prior modeling studies vary from 0.3 to 6 percent/hour (Newman 1980b). The large range of values for this parameter reflects the uncertainty in representing the various chemical pathways for conversion. Currently, the sulfate formation rate cannot be calculated from fundamental reaction mechanisms because the rate-controlling processes have not been fully clarified. Therefore, rates must be chosen from among those measured (or inferred from measurements) under conditions as close as possible to those of interest.

A comprehensive review of current long-range transport models has been presented by Eliassen (1980). Many of the models described in that review have been applied and their predictions evaluated against data. EPA's proposal of at least one long-range transport model appropriate for predicting pollutant concentrations over distances of 100 to 1,000 km from their source is a welcomed development.

Accuracy and Comparability of Models

The accuracy of a calculated concentration is affected by uncertainties in input data--such as the emission inventory and meteorological data--and the degree of applicability of the model to the situation at hand. Models can be used more accurately to examine differences between alternatives than as means to calculate an absolute concentration.

Because of uncertainties in the meteorological fields used as input, models generally predict the magnitude of a concentration at a given point more accurately than they predict the precise location where a concentration of given magnitude will occur. Also, the accuracy of predicting the area containing all points where concentrations are above a certain value is usually higher than that of predicting the concentration at a particular point. Finally, the accuracy of calculated long-term (e.g., annual) average concentrations is better than that of maximum 1-, 3-, or 24-hour concentrations, because mean values are more accurately predicted than extreme values.

If an applicant does not want to use a model recommended by EPA, approval may be sought to employ another model by demonstrating that the alternative is "comparable" to a recommended model. Comparability is judged on the degree to which the alternative model treats individual features of the modeling problem in a manner similar to that of the reference model. The essential problem with this feature of the regulations is not the basis for judging comparability but the fact that the models to which the alternative model is to be shown comparable are themselves often inappropriate for the situation being modeled. Thus, if a Gaussian model is invalid under a particular set of conditions, use of a model based on a comparable set of assumptions cannot be expected to produce any more valid predictions.

Comparability is a useful concept to employ when applicants request use of alternative models. But in judging comparability the key problem remains--i.e., that the EPA-approved models must be upgraded and revised to represent more adequately the physical phenomena of pollutant transport and dispersion.

Validation of models entails repeated tests of performance for the variety of conditions to which each model applies, leading to confidence in the results of future trials. Tests to validate models require good meteorological data and accurate monitoring of test species. Because of restrictions placed on calculations in preconstruction review (for example, the value for stack height under theoretical good engineering practice must be used rather than the actual height) and because of the difficulties of measuring small changes in ambient concentrations, there are only limited opportunities to validate models by using monitoring data and modeling results obtained to support permit applications. Validation of models for use in the PSD process may thus require special tests. Models for predicting worst-case conditions are inherently more difficult to validate than models for statistical parameters.

MODELING AND A STATISTICAL APPROACH TO MANAGING AIR QUALITY

Atmospheric concentrations of pollutants vary from hour to hour and day to day, and changes over short periods are quite difficult to predict. The behavior of the atmosphere over long periods is more regular and predictable, and thus average concentrations are easier to predict provided emissions are predictable. For example, a forecast of the maximum ambient concentration of a pollutant, say SO_2, on a given day is generally less reliable than a prediction of how many days in a year the concentration will be equal to or greater than a given value. The behavior of the atmosphere can be described statistically; and taking atmospheric statistics into account can have important consequences for the PSD program.

Distributions of Concentrations of Pollutants

The statistical properties of the atmosphere can be described by distribution functions. To obtain a distribution function of, for instance, the hourly concentration of SO_2, one measures the hourly concentrations over an extended period and plots the frequency of occurrence of concentrations as a function of concentration. The result is a histogram of the type illustrated in Figure 3.1.

If many measurements are taken for many values of the concentrations, the histogram can be replaced by a smooth curve, as shown in Figure 3.1. Notice that low and very high concentrations occur only rarely and that the concentration that occurs most frequently need not be the average or mean concentration. The width or variance of a distribution such as that of Figure 3.1 is a measure of how variable the concentration is from day to day.

While there is no a priori reason to expect that air quality distributions would have analytic mathematical representations, it turns out that many do--or at least that they are so close to the plot of an equation that the equation can be used as a valid approximation. Many air quality data, when plotted against the logarithms of the concentration rather than against the concentration, yield plots that are Gaussian or normal (i.e., bell shaped). A distribution with this property is called a log-normal distribution. Log-normal distributions have been widely used to represent air pollution concentration data (Bencala and Seinfeld 1976, Larsen 1971).

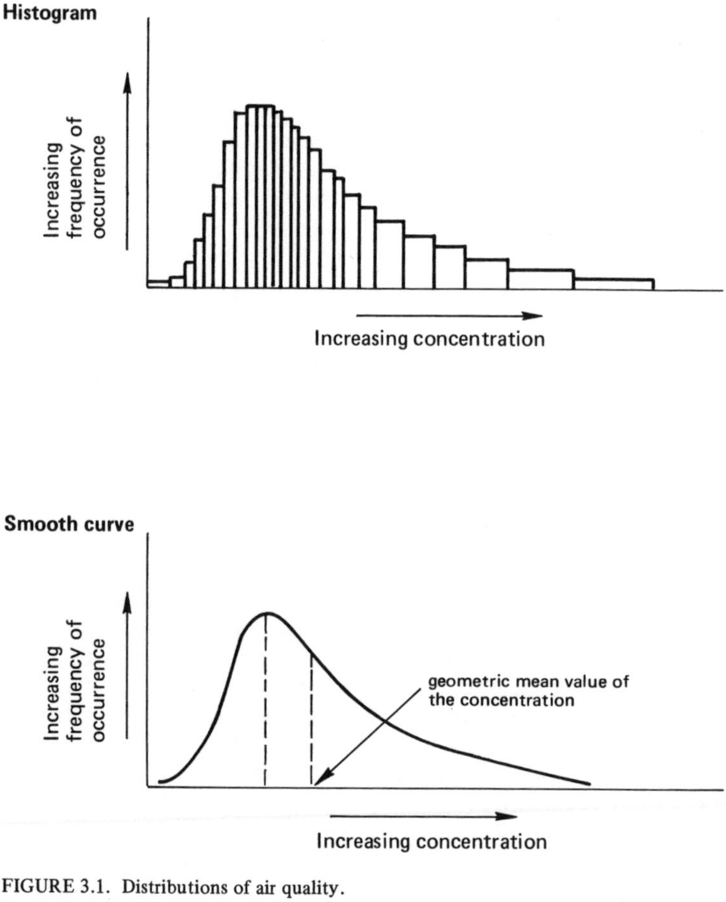

FIGURE 3.1. Distributions of air quality.

The log-normal distribution has the helpful characteristic that, when plotted on special graph paper, the curve becomes a straight line. The graph paper has a logarithmic ordinate for the concentration and a probability scale on the abscissa for plotting the cumulative frequency of occurrence, i.e., the total frequency of occurrence of all concentrations greater than each point on the plot. Figure 3.2 shows the distribution of 1-hour average SO_2 concentrations equal to or in excess of the stated values for Washington, D.C., for the seven-year period from December 1, 1961 to

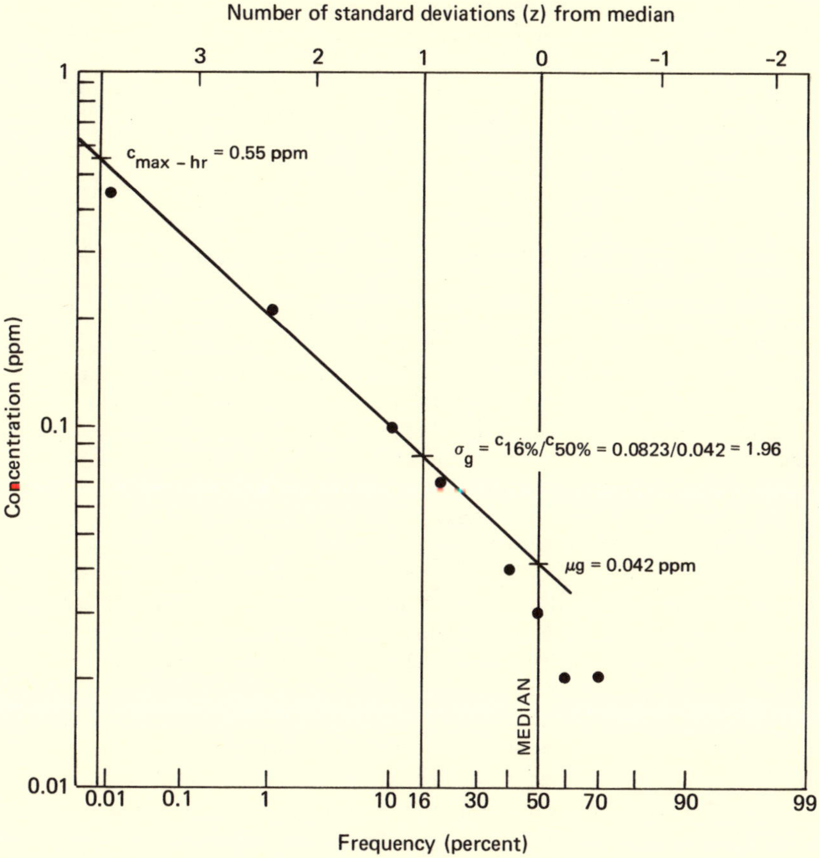

SOURCE: Larsen (1971).

FIGURE 3.2. Frequency of 1-hour average sulfur dioxide concentrations equal to or in excess of stated values for Washington, D.C., December 1, 1961-December 1, 1968.

December 1, 1968 (Larsen 1971). From this plot we see that half the readings were above 0.042 parts per million (ppm) and 16 percent were above 0.081 ppm.

The log-normal distribution is completely characterized by two parameters, the geometric mean μ_g and the standard geometric deviation σ_g. The geometric mean or median is the concentration where the straight line plot crosses the 50th percentile. The slope of the line is related to the standard geometric deviation, which can be calculated from the plot by dividing the 16th percentile concentration (which is one standard deviation from the mean) by the 50th percentile concentration (the geometric mean). For the distribution of Figure 3.2, μ_g = 0.042 ppm and σ_g = 1.96. Plots such as Figure 3.2 are widely used in air quality analysis, because it is important to know how often concentrations equal or exceed certain values--such as the NAAQS or the PSD baseline plus increments.

Predicting Ambient Concentrations

Ambient standards and the maximum allowable increases under PSD are stated in terms of a maximum concentration for an averaging time with a limit to the number of times per year that the maximum concentration can be exceeded. For example, the primary NAAQS for maximum concentration of TSP in a 24-hour period is 260 $\mu g/m^3$, not to be exceeded more than once per year. The PSD increments are similarly expressed. Continuous sampling data for all regulated pollutants are rarely available, and all we frequently have are periodic samples taken over relatively short averaging times. Compliance with standards must be assessed with incomplete records of ambient conditions, but this assessment can be made by taking a statistical approach. If sufficient data are available to determine that the distribution of concentrations is analytic--or if there is sufficient reason to assume a specific distribution--it is possible to calculate the probability that a given level will be exceeded in one or more periods of time.

To predict extreme concentrations, we must be able to relate the distribution of future concentrations to current ones given assumptions about increases or decreases in emissions. For example, if future emissions were to increase to twice their current values, we would want to find out whether extreme concentrations at distant sites would double or increase by more or less than that amount. If today's emission rate were multiplied by a positive factor κ, if background concentrations were negligible, and if meteorological effects remained unchanged, the expected total amount of pollutants affecting a given site over the same period would also be multiplied by the constant κ. Larson (1971) and others have observed that if meteorological and terrain conditions remain unchanged, the standard geometric deviation of the log-normal distributions of pollutant concentrations will also remain unchanged. If the assumption is valid and log-normal distributions are applicable, we can calculate the distributions of future concentrations knowing current distributions and changes in emission rates.

Figure 3.3 shows two log-normal distributions with the same standard geometric deviation but different geometric mean values. Let us suppose that distribution #1 represents current conditions and therefore that the current probability of exceeding a concentration of about

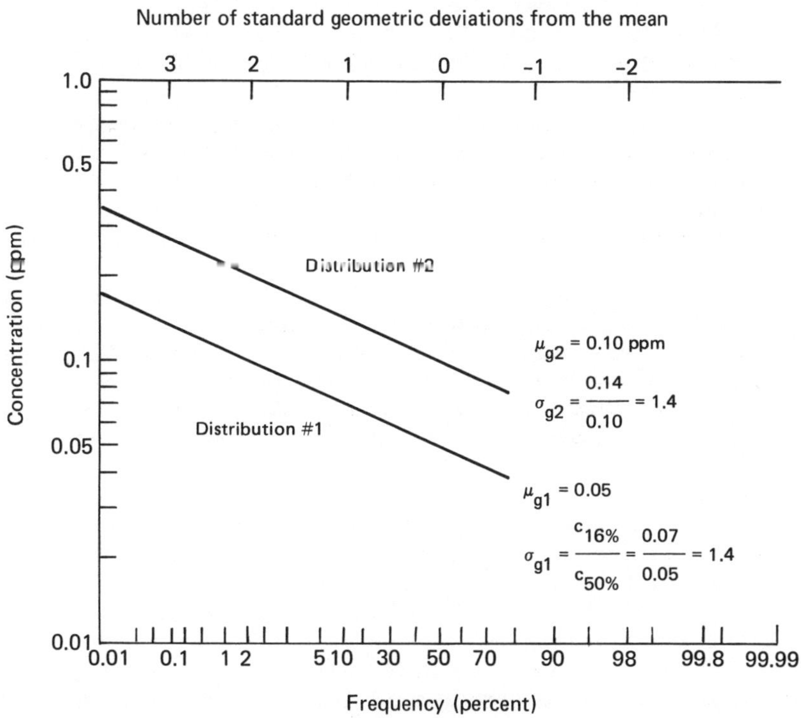

FIGURE 3.3. Two log-normal distributions with the same standard geometric deviation.

0.13 ppm is about 0.0027--which corresponds to about one day per year if the distribution is of 24-hour averages. If the emission rate were doubled, the new distribution function would be given by distribution #2. The new distribution has a median value twice that of the old one since total loadings due to emissions have doubled. Under the new case, a concentration of 0.13 ppm will be exceeded 22 percent of the time, or about 80 days a year. Under the new case, the concentration that is exceeded only one day per year rises to 0.26 ppm.

Statistics and PSD Increments

The increments for SO_2 and TSP are stated for three averaging times: 3-hour and 24-hour maxima, and an annual average (Table 1.2). If the three increment levels for each pollutant were consistent with the same (log-normal) distribution, no one of the three would be a more severe constraint than the others. There is no evidence, however, that the statutory increments for either SO_2 or TSP were selected with statistical distributions in mind. Therefore it is likely that in a specific PSD case, one of the three increments for each pollutant will be the constraining one.

While one averaging time or another may be important in a given case for protecting AQVRs in a Class I area, there appears to be no reason for maintaining sets of inconsistent increments for Class II or Class III areas--where general air quality is the primary concern. If a set of increments for a pollutant is consistent with a distribution, consideration of all the increments for a pollutant can be replaced with consideration of only the distribution and one averaging time. It should therefore be practical to simplify the PSD increments system in Class II and Class III areas by measuring--or assuming--the form of the distribution of concentrations of a pollutant and selecting a value for an increment at one averaging time; this simplification can be accomplished without changing the degree of protection provided by the current system. If different averaging times are not required to protect different AQRVs in Class I areas, this simplification could be extended to Class I areas as well.

Currently, one of the difficulties encountered in implementing the PSD program is in modeling for rare events. A maximum 24-hour average concentration that can be exceeded only once a year has a probability of being exceeded in a year of less than 0.003; a maximum 3-hour concentration that can be exceeded only once per year has a probability of being exceeded in a year of about 0.00035. Thus, exceeding the maximum is a rare event. As explained earlier in this chapter, the rarer the event, the greater the uncertainty in the results of modeling.

Increments based on short (3-hour or 24-hour) averaging times require modeling schemes different from those required by a yearly or seasonal average. When a 3- or 24-hour value is to be calculated, the daily set of meteorological conditions most conducive to high concentrations is determined and the model is worked for that set of conditions. Because of the different ways in which different models treat meteorological variables, model predictions of short-averaging time increment consumption may differ considerably. When dealing with the annual average increment, models generally calculate concentrations for every day of the year and average the results over the entire year. As a result, differences in model formulations tend to be averaged out, and the annual average concentrations predicted by several models will usually agree rather closely. Therefore, the use of long-term average increments has the effect of reducing the importance of differences among models.

A disadvantage of focusing on long-term averages is that the absolute values of the modeled concentrations are smaller, which may require greater precision (a larger number of significant digits in the results) and make it all the more difficult to monitor for compliance. Further study is needed to determine whether the advantages of simplifying the increment system by reducing the number of increments and focusing on long-term averages outweigh the disadvantages.

Statistics and Compliance

Another alternative that would bring the PSD system into closer conformity with the capabilities of modeling would be to change the criteria for compliance. Currently the PSD increments--and the NAAQS--may be exceeded only once per year. Modeling the extreme occurrence of short-term maxima, as indicated earlier, is difficult; the results are uncertain and arguable.

Greater confidence in the results of modeling might be achieved with a change in the criteria for compliance. For example, if instead of only once per year, increments or standards could be exceeded five percent of the time, the statistical reliability of modeling would be greatly enhanced. Making such a change in compliance criteria need not result in altered levels of protection, because the numerical values of increments or standards could be adjusted, based on knowledge of distributions, so that the new values reflect the same level of air quality as the old. If 95 percent of observations of a 24-hour maximum concentration were required to be below a certain value, that value could be exceeded 18 days per year. It is to be expected that predictions using different models would not vary substantially from one another on such a basis. The World Health Organization (1972) has recommended that it be permissible for pollutant concentrations to exceed a standard for two percent of observations provided that no two such events occur on consecutive days.

Such a change in measures of compliance could have far-reaching effects on all air quality programs. Therefore we believe that further study is needed to develop definitive conclusions and recommendations. We also believe that the current criteria for compliance should be subject to scrutiny and review. We recommend that studies be conducted to determine the advantages and disadvantages of alternative expressions of the criteria for compliance for the PSD program as well as for other programs of the Clean Air Act.

MODELING DEGRADATION OF AQRVs

To assess the potential effects of proposed emissions on AQRVs--as required at least for Class I areas containing federal lands--models relating emissions to changes in AQRVs must be used. These models consist, in general, of three parts:

1. quantification of background conditions (such as background pollutant concentrations or status of the AQRV);
2. calculation of the dispersion and transformation of emissions (including diffusion, transport, formation of secondary aerosols and removal rates); and

3. estimation of the effects of changes in concentration on the AQRV (including the human perception of these effects).

The first two parts constitute models for air pollutant concentrations; the third part adds the component specific to the AQRV of interest. Each part of the model involves limitations and potential errors, leading to significant uncertainties in the overall modeling results. In the case of visibility, prediction of optical effects is one of the least uncertain, from which we can infer that visibility modeling is not significantly less accurate than modeling only for pollutant concentrations. Modeling human perceptions of changes in optical properties of the atmosphere, however, is uncertain.

Visibility

There are two basic types of visibility models: deterministic and empirical. Deterministic models involve the solution of mathematical equations describing the physics, chemistry, and optics of atmospheric visibility phenomena. Empirical models attempt to make predictions based on the analysis of actual monitoring data. Because air quality modeling is uncertain, it is often advisable to use both the deterministic and empirical approaches and to compare results. Both types of models can be formulated for either local plume blight problems or large-scale regional haze problems.

There has been considerable effort recently to develop deterministic visibility models (Latimer et al. 1978, U.S. EPA 1979a). The greatest part of this effort has been directed toward predictions of plume blight; regional haze models are in the earliest stages of development. EPA has recently issued a user's manual for a specific model (Johnson et al. 1980) and a workbook for estimating degradation of visibility (Latimer and Ireson 1980). Because deterministic visibility models are based on calculations of ambient concentrations using Gaussian air quality models, they are subject to the same limitations as Gaussian models. In addition, there are considerable uncertainties in the estimation of effects on human perceptions (Malm et al. 1979a); few data on the relationships between atmospheric conditions and perceptions of degradation of visibility are available on which to base projections.

Empirical models for plume blight can be based on field study data concerning plumes from power plants, copper smelters, and other sources (Macias et al. 1979, Ogren et al. 1976). When a single pollutant such as NO_2 dominates, the models are very simple. The pollutant concentrations, light scattering, and absorption can be factored by the total emission rate of the pollutant. When several pollutants are involved, special calculations must often be performed to determine the individual effect of each pollutant in the field study plume. Once these individual effects are estimated, the light scattering and absorption from each individual pollutant is factored by the appropriate emission rate.

An empirical model relating regional haze to sulfur oxide emissions in the Southwest has been formulated by Marians and Trijonis (1979). Using visibility and emission data for the 1967-1968 copper smelter strike as well as for the entire period 1949-1976, the researchers estimated coefficients relating regional extinction due to sulfates to the SO_2 emission rates. Separate coefficients were derived for the effects of SO_2 emissions on three spatial scales; less than 50 miles, 50 to 200 miles, and 200 to 500 miles from the source. These coefficients permit a rudimentary analysis of the regional haze caused by a given amount of SO_2 emission in the Southwest.

Empirical analyses can also be used to help estimate parameters needed in deterministic models. For example, many researchers have analyzed air quality data to determine the extinction efficiency of sulfates as a function of relative humidity (Cass 1976; Leaderer and Stolwijk 1979; Trijonis 1979, 1980; Trijonis and Yuan 1978; White and Roberts 1977). These empirical coefficients can be used to determine optical effects from pollutant concentrations computed with deterministic models (Latimer et al. 1978).

Acid Precipitation

Models of acid precipitation must include descriptions of long-range transport processes, the processes of physical and chemical transformation of species containing sulfur and nitrogen, and the physics and chemistry of precipitation. Cloud formation and precipitation processes must also be represented in specifying the manner of incorporation of the gaseous species into rainwater or snow.

There are two approaches to modeling the chemistry of precipitation. One approach uses a material balance that integrates over an entire storm; this technique ignores the chemistry of the aqueous phase (Dana et al. 1975, Hales 1972, Hales and Dana 1979a,b). The product is a scavenging ratio defined as the ratio of the concentration of the pollutant in water to that in air. Scavenging ratios are simple to use in predicting acidity of rainfall. A second approach uses a material balance in a small volume of the atmosphere in a storm or cloud and calculates changes in concentrations in the volume and flows across its surface (Hales et al. 1973, Scott 1978). This differential material balance uses analytical or numerical models of the coupled continuity equations for the gaseous and liquid phases.

Most models of acid precipitation currently available are of limited usefulness. One reason is that models are limited to one or two chemical species whereas many species should be included in a material balance if the chemistry of the hydrogen ion is to be adequately represented. Furthermore, there is an inadequate understanding of mechanisms of interphase transport and of the flow fields in clouds. Finally, as we have noted, scavenging models, which are easy to apply, ignore aqueous-phase chemistry. These shortcomings make development of improved capabilities in modeling acid precipitation an important research need (Henderson et al. 1980).

Odor

Odorants disperse rapidly in the atmosphere by turbulent diffusion; they also undergo complex physical and chemical reactions in the air. To date, attempts to model odors have been limited to short distances from the source and to nonreactive pollutants (NRC 1979c). Models for odor, like those of visibility, must consist of essentially three parts: an assessment of background conditions, a calculation of changes in ambient concentrations of the pollutants of interest, and an assessment of the perception of the effects of those changes. As in the case of visibility, the weakest link in modeling odors is in predicting the human sensory response.

4
SITING AND INDUSTRIAL DEVELOPMENT

The PSD provisions in the Clean Air Act limit the availability of air as a resource for industrial development. For any given state of technology, then, the time may come when continued realization of the goals of the Act can be achieved only with greatly increased costs for further industrial development; without some relief, consumption of increments in all classes of attainment areas will limit construction of new sources. It is also clear that by designating areas for special protection, as in the case of Class I areas, costs of acquiring desirable sites, obtaining permits, and controlling emissions will likely increase, perhaps resulting in modification, relocation, or foregoing of industrial development plans.

It is extremely difficult, however, to project the extent or value of changes in development plans nationwide or to forecast in advance what sites will be made economically unattractive because of PSD constraints. The legislative history of the 1977 amendments makes clear that PSD increments were not intended as absolute standards for determining the proper balance between clean air and economic development, and one of the difficulties in making such estimates is that the PSD program allows many variations in application.

Thus, early attempts to project buffer zones around Class I areas on the assumption that increments were absolute standards were explicitly rebuffed by Congress, as explained by then Senator Muskie (1977, p. 9169):

> Much confusion has been generated about so-called buffer zones that encircle the Class I regions. This

> bill eliminates the infexibility of current EPA regulations by establishing the Class I increments solely as a means of determining where the burden of proof should lie as to adverse impact on air quality values. Like the Class II numbers, the Class I increments are an index of the change in air quality. They do not, in any way, establish a final basis for approval or disapproval of a permit application. Thus, any maps which describe buffer zones as a result of the Senate bill distort the impact of these amendments.

Because the PSD approach requires "a balancing judgment to be made on a case-by-case basis" (Muskie 1977, p. 9170), it is impossible to describe the economic impact of PSD except in a very general way. Even if it turns out that the waiver, variance, and redesignation procedures currently in the program are not practical in application, variations inherent in the PSD program make it difficult to assess the economic consequences of constraints on development. Among these variations, the most obvious include:

Variations in administration by states. The state role in implementing PSD is more prominent than it is under other parts of the Clean Air Act. Like primary and secondary standards, which are enforced through federally approved but state-designed implementation plans, PSD provisions are also open to state enforcement through approved permit programs. But because the PSD increments, unlike primary and secondary standards, are not intended to be rigid constraints, the state has an additional role to play in the process of determining whether permits will be issued in cases where AQRVs may not be threatened, even though increments have been consumed. States may also adopt different procedures for allocating available increments to applicants. Finally, states may exhibit various degrees of innovation and initiative in seeking ways of meeting both PSD and developmental objectives.

The economic climate in the state and the general attitude toward industrial development are also potentially important factors in determining how PSD provisions affect development in practice. When the 1977 amendments were enacted it was already clear, for example, that the "the electric utility industry likely will be affected more than any other industry by prevention of significant deterioration requirements" (U.S. House 1977, p. 155). In view of this potential impact, rate-setting authorities should be aware that varying degrees of sympathy for the attempts of utilities to include the costs of pollution control in their rates will result in varying degrees of impact of PSD on the industry as a whole.

Variations in resources for development. The potential impact on siting and development decisions appears to be greatest where protected Class I areas and valuable, accessible resources are in close proximity and where large mineral deposits are found in small geographic areas, in which case intense development might be constrained by Class II increments. As new resources are discovered, or alternative means of extracting resources are developed, the economic impact of PSD will also change.

Other site-specific factors. Among the variables that determine the actual impact of PSD on siting and development, local meteorological conditions are the most obvious. For example, meteorological conditions and terrain might allow comparable facilities to be located much closer to a Class I area in one direction than in another. Other variables include differences in adaptability of control technology to the location in question, differences in the extent to which various industries can afford control technology, and differences in industry's ability to relocate proposed facilities to environmentally less sensitive sites.

The number, distribution, and size of Class I areas relative to Class II areas also differ from region to region, with consequent differences in impact on development. Variations in emission standards and air quality among the states will also complicate the task of attributing siting and development problems specifically to PSD.

CONSTRAINTS ON SITING

A number of important factors, including effects on air quality, must be considered in deciding where to build a major new industrial facility in a PSD area. Each of these factors may in turn be subject to its own specific regulatory controls. The process of selecting a site is a complex one, involving many compromises. Among the factors to be considered are availability of land, water, and transportation; proximity to energy or mineral resources and to markets; means for disposal of wastes; hydrology, geology, and seismology; and public acceptance.

Local and state-imposed constraints on land use must be considered and, if federal action is involved, alternative sites must be examined as required by the National Environmental Policy Act. Alternatives to be considered typically include existing sites where expansion is possible; other company-owned land suitable for development; and land owned by others available for acquisition and suitable for development as a plant site. The desirability of alternative sites may depend strongly on proximity to resources. Suitability of soils, potential for seismic disturbances, and proximity to other land uses also affect the availability of land for development.

The availability of potable and process water, access to the relatively large amounts of water required for condenser cooling systems, and the potential for site flooding are all examples of considerations that can affect the siting decision. Consideration must also be given to regulatory constraints on the consumptive use of water, compliance with State Water Quality Standards, and limitations on water use imposed by minimum streamflow requirements.

Adequate and reliable railroad, highway, or waterway transportation facilities must exist for delivery of fuel, removal of waste materials, and shipment of products. The greater the distance from existing distribution networks, the greater the cost of transporting or distributing the final product.

The manner and place of disposal of wastes must be considered, including restrictions of the Clean Water Act on discharge to waters and of the Resource Conservation and Recovery Act on the treatment of hazardous and nonhazardous wastes. Siting decisions must take account of meteorology, terrain, and consequences of air emissions on the NAAQS. In addition, effects on sensitive ecosystems and endangered species must be considered. And sometimes potential environmental risks to structures from, for example, violent storms or seismic activity affect both siting and design decisions.

The combination of local meteorological conditions near Class I areas, terrain, and the number and kinds of other PSD sources will limit the availability of sites for development and ultimately determine the impact of PSD on particular proposed facilities. Thus, in addition to the usual factors affecting siting decisions, applicants for PSD permits must consider not only location with respect to Class I areas, but also the meteorological conditions that affect the fate of emissions from a proposed site. Proximity to other sources and the combined effects of emissions from them on both distant Class I areas and on nearby attainment areas (Class II or Class III) must also be considered. The effects of emissions on the status of increments (and NAAQS) must be modeled.

Another consideration is the emission limitation imposed by the required use of best available control technology. If the technology for achieving the required levels of control entails unacceptable costs and if the applicant does not wish to abandon the project, the applicant must seek a waiver or variance (if Class I increments are the cnstraining factor), find an alternative site where emission limitations are less costly, pursue a reclassificaton of the area, or obtain compensating reductions in emissions (offsets) from existing sources.

Making the Siting Decision

This survey of factors relevant to siting decisions makes it clear that the actual choice of site requires complex tradeoffs of often competing factors, most of which are economic and some of which are environmental. A typical example is that of a coal-fired power plant using lignite, to be located in the Northern Great Plains. As a general rule, the most economical plant would be located at the source of the coal, since it is uneconomical to ship lignite by rail more than about 100 miles. If rail shipment is required, the plant must be close to that portion of the existing rail system capable of carrying unit trains. It is also desirable to locate near existing transmission lines. Finally, availability of water is a constraint in many places. Added to these are the environmental constraints, including air quality, water quality, waste disposal, and the potential impact on PSD areas. The additional costs of BACT--over and above those of NSPS--for each potential site must be weighed against other site-specific costs or savings before a firm can decide what to do. The flexibility provided in BACT determinations and the waivers and variances--hence in air pollution control costs--further complicate the business analysis.

PSD as a Formal Mechanism for Making These Tradeoffs

The process of applying for PSD permits and modeling air quality effects for alternative sites thus is intended to force judgments about how to balance these environmental and economic concerns in the siting decision. State control of this process through the PSD program incorporates public concern for the importance of both air quality and economic development into the siting decision. This process inevitably entails additional costs for the developer, not only in control technology but also in time delays. The process may also result in the choice of a site that is less economical in terms, for example, of the cost of transporting water or fuel.

One role of the permitting authority in making BACT determinations is to include economic impacts in addition to the extra costs to operators of pollution control in the balancing of economic and environmental concerns. For example, benefits may accrue to local economies from stringent emission limitations that allow additional firms to have access to portions of increments that are thereby conserved. The consumers of products and services provided by these firms will pay higher prices. In our discussions with practioners, we found that assessment of these potential impacts, which again can only be made on a case-by-case basis, is almost universally overlooked.

PSD, RESOURCE MANAGEMENT, AND INDUSTRIAL DEVELOPMENT

The purpose of PSD is not to foreclose development, but to manage economic growth in a manner consistent with concerns about good air quality. As far as we are aware, up to the summer of 1980 no application for development in a PSD area had been rejected for inability to comply with PSD, though some proposed facilities had been modified or relocated. The BACT requirement is intended to ensure that each PSD source uses as little of the air resource as is feasible, leaving as much of that resource as possible for subsequent use by others.

Availability of Increment as a Critical Factor

Operation of PSD Where Increments Have Not Been Consumed

As other chapters of this report indicate, the PSD increments are legislated constructs, intended to alert decision makers to look for actual effects or to examine the relative values of air quality and further economic development in appropriate cases. In cases affecting Class I areas, where models predict that an increment has been or will be consumed, the focus shifts from the increments to an actual effects-based assessment of proposed development. When Class II increments are threatened with being exceeded, the focus shifts to consideration of redesignation.

In theory, permits can be denied despite available increment if an effects-based assessment indicates that a proposed development will adversely affect AQRVs in Class I areas containing federal lands. The burden of proof of such an adverse effect is on the federal land manager, and to our knowledge, no such showing--indeed no attempt at such a showing--has yet been made in the operation of the Act. In general, then, as long as increments are available, applicants under a first-come, first-served allocation scheme may expect to obtain PSD permits if they are willing to meet emission limitations that are imposed.

Waiver and Variance Procedures When Class I Increments Have Been Consumed

Only in a limited number of cases have states been faced with applications for permits from sources whose emissions have been predicted to exceed the available PSD increments. No permits have actually been denied to date, although in some instances, applicants have chosen alternative sites where emissions would not cause increments to be exceeded. Examples of such relocation include the IPP project in Utah and the Western Gas Processors Plant in North Dakota. The imminent prospect of increment consumption, however, raises the question of how and whether future development can proceed if relocation is not practical.

In this case, too, the Act and legislative history make clear that consumption of Class I increments is meant to trigger a more detailed effects-based analysis of the impact of proposed development on AQRVs (U.S. Senate 1977, p. 35). Now, however, the burden of proof is on the applicant to satisfy the federal land manager that no damage to the values of the Class I area will occur. If successful in making this showing, the applicant may be granted a waiver requiring the facility not to exceed new increments identical to the Class II allowable increases, with the exception of the 3-hour SO_2 maximum (see Table 1.4). The increases in pollutant concentrations allowable under this waiver procedure are from four to twenty times greater than the small increments permitted under Class I restrictions.

Despite the apparent Congressional intent that the Class I increment and waiver provisions were to be part of a single permit process, it appears in practice that the increments are given considerably greater--indeed almost controlling weight--in the review process. Several reasons appear to account for this result. First, it has not generally been necessary to pursue the waiver provision because in most cases increments have not yet been consumed. Second, given the apparent readiness to grant permits as long as increments have been available, the focus of most activity has been on refining models to show that emissions have not caused increments to be exceeded--i.e., that increment is still available. Finally, and most likely, the subjective nature of the decision that is required of federal land managers (who have an affirmative responsiblity to protect AQRVs) under the waiver provision has a clearly discouraging effect on industry's willingness to pursue this possibility.

We are aware of one case in which the possibility of a waiver is under consideration (L. Marshall, U.S. EPA, Region X, Seattle, private communication, January 28, 1981). The Northern Tier Pipeline Company has applied for a PSD permit for operation of a terminal at Port Angeles, Washington. Port operations would cause Class I increments to be exceeded in a narrow strip of land in Olympic National Park. An analysis of the potential effects of operations at the proposed terminal on visibility, soil, and vegetation has been presented to the National Park Service for consideration.

This observation deserves emphasis: when increments are consumed, the likelihood that development will be constrained despite the theoretical availability of a Class I waiver from the federal land manager is a very real one as long as no clear standards govern the effects-based determination. In this respect, improved understanding of AQRVs and the amounts and kinds of pollutants that affect them (see Chapter 2) are critical to ensuring that the intended balance between environmental and developmental concerns is reached.

In addition to the general Class I waiver provision, the PSD section also provides for a variance from the 3-hour and 24-hour SO_2 increments in certain cases. The SO_2 variance provision (see Table 1.5) differs from the more general waiver in several respects. First, the applicant's showing that AQRVs will remain unimpaired is made to the governor of the state, rather than to the federal land manager. In the event of disagreement between the federal land manager and the governor, the matter is sent to the President, who decides on the basis of the national interest whether to grant the variance. The President's decision is declared not to be reviewable in any court. Second, maximum increases allowed under the SO_2 variance are considerably more stringent than those available under the waiver provision. In high terrain areas, the SO_2 variance increments are only equal to about one-third of the increments available under the waiver approved by the federal land manager.

As far as we are aware, there are no instances of applicants seeking SO_2 variances. The potential advantage of applying to the state rather than to the federal land manager may well be offset by the more cumbersome procedures and the more stringent limitations of the SO_2 variance.

Reclassification

Another potential source of flexibility to accommodate development where PSD increments are threatened is reclassification. Except for mandatory Class I areas, states are given wide discretion to reclassify. For example, Class II areas may be redesignated either Class I or Class III. A state may redesignate any area (that is not a mandatory Class I area) to Class III if certain procedures are followed, including notice and opportunity for public hearing and comment and a detailed assessment of the environmental, economic, and social effects of redesignation. In those cases where the constraints of mandatory Class I increments are the limiting factor, these provisions for reclassification will not provide a means for choosing development over the air quality goals of the Act. Where Class II constraints are critical, these provisions are supposed to give control over the balance between economic growth and environmental protection to the states. Redesignation of lands from Class II to Class III may, however, be impractical as a means for reaching local decisions on the relative importance of economic development and protection of air quality, because the redesignation process is so complex. Here again, it is too early to tell; the first such application is under consideration (Illinois EPA 1980).

Compensating Reductions in Existing Emissions

When increments in an area have been consumed and neither variances, waivers, alternative sites, nor reclassification are practical options, an applicant still might be able to proceed with a project provided the emissions of the proposed new source were offset by reductions in emissions from existing sources. In some cases an operator might reduce emissions from his own existing plant to facilitate an expansion. Alternatively, reductions might be achieved either by asking the state authority to require existing sources to reduce emissions through revision of the SIP or by financing reductions in emissions from one or more existing plants (that is, buying offsets from other operators). In either case, new, more stringent controls would have to be applied to existing equipment. Retrofitting generally results in higher control costs than if the more stringent emission limitations were specified in the original design.

The potential for the use of offsets may be limited in PSD areas for several reasons. First, there are likely to be few sources in any given PSD area, a fact that limits flexibility. Second, the major sources from which offsets might be obtained are likely to be large, so that if substantial reductions are required, applying additional controls to them could be expensive. Smelters, however, are an exception. They make major contributions to regional emissions of SO_2 in the West and costs of retrofitting SO_2 controls on smelters per ton of SO_2 removed are lower than costs of scrubbers on new electric generating plants. Finally, many operators for whom reductions in emissions are technically and economically practical might want to retain the potential of getting an offset to accommodate their own future expansion and hence might be unwilling to cooperate with new developers.

ESTIMATES OF IMPACTS ON SITING AND DEVELOPMENT

Because of the inherent problems in making a realistic generic assessment for the relatively short time that the PSD program has been in operation, and the limited resources and expertise available to us, we have not attempted an overall assessment of the economic costs involved in implementing PSD.

Estimates of the economic impact of PSD were considered by Congress when the 1977 Amendments were passed. The most frequently cited analysis was that prepared jointly by the Federal Energy Administration and the U.S. Environmental Protection Agency (U.S. House 1977). This analysis focused on the electric utility industry and the potential impact of the proposed PSD program on the construction and operation of large coal-fired electric generating plants. The analysis at that time concluded generally that proposed Class II and Class III increments would not affect the siting of such facilities at all and that Class I areas would not significantly affect the siting of even the largest of such sources.

The 1977 amendments changed certain critical factors on which the EPA-FEA report was based; the increments, for example, were made more stringent. Subsequent estimates, prepared since the PSD provisions were enacted, include two separate reports prepared for EPA (ICF 1980, Latimer 1979). The purpose of the ICF study was to provide a preliminary indication of the economic impacts of possible visibility regulations applicable to existing major stationary sources. Alternative assumptions were made concerning both the sources likely to be affected and the stringency of the visibility regulations that might be imposed. Cost estimates were made for various combinations of assumptions. Under stringent regulatory assumptions, estimated impacts on utility rates on specific plants ranged up to 3.1 percent under one set of criteria for selecting the potentially affected plants and up to 32.9 percent under an alternative set. Under less stringent assumptions about what constitutes unacceptable degradation, rates were estimated to increase by up to 2.6 percent or 19.4

percent, depending on the criteria for selecting the potentially affected plants.

The Latimer report bears more directly on the impact of PSD. The report analyzes the impact of power plants of various sizes, which emit SO_2 and NO_x at various rates, using methods of analysis that have since been incorporated by EPA into its draft recommendations of methods for estimating visibility impairment caused by plumes (Johnson et al. 1980, Latimer and Ireson 1980). Latimer concluded that the most restrictive limitations on siting power plants were the Class I increments for 3-hour and 24-hour maximum concentrations of SO_2; these increments could restrict siting of large coal-fired plants located in Class II areas as far as 100 to 200 km from Class I areas. He also calculated reductions in visual range caused by emissions from a 2,000 megawatt (MWe) power plant under worst-case meteorological conditions. The results of some of his calculations are shown in Table 4.1. Ground-level concentrations and percentage reductions in visual range are shown at a distance of 350 km from the source, assuming a conversion rate from SO_2 to sulfate of 0.5 percent per hour. The table shows results for an observer sighting along and across the plume and for two values of background visual range. The background visual range of 130 km is typical of the western United States; that of 15 km is more typical of the East. Reduction in visual range during worst-case meteorological conditions is larger by at least a factor of five than it would be under average

TABLE 4.1. Effects of a hypothetical power plant on distant air quality.*

Emission rate ($lb/10^6$ Btu)	Ground-level SO_2 concentration ($\mu g/m^3$)	Percent reduction in visual range: Sighting along plume VR_0=130 km	Sighting along plume VR_0=15 km	Sighting across plume VR_0=130 km
0.2	1	11	2	4
0.5	3	28	4	9
0.8	5	40	7	14
1.2	8	51	11	21

*Effects were calculated at a distance of 350 km from a 2,000 MWe plant in level terrain under worst-case meteorological conditions with a conversion rate for SO_2 to sulfate of 0.5 percent per hour.

SOURCE: Latimer (1979).

meteorological conditions with higher wind speeds and greater mixing depths.

Latimer reached the following tentative conclusions about constraints on power-plant siting:

- Power plants that are 1,000 MWe or larger emitting SO_2 at a rate of 340 ng/J (0.8 lb/10^6 Btu), or 2,000 MWe or larger emitting 210 ng/J (0.5 lb/10^6 Btu), may cause maximum 24-hour-average ground-level SO_2 concentrations in excess of the PSD Class II increment of 91 $\mu g/m^3$, according to calculations based on the conservative screening techniques recommended by the EPA.
- If the VALLEY model is used as a basis for evaluation, 2,000 MWe power plants will have to be sited at least 18 km away from elevated terrain if SO_2 emissions are well controlled (less than 86 ng/J or 0.2 lb/10^6 Btu) and at least 50 km away if SO_2 emissions are 340 ng/J in order to meet Class II PSD increments. Even small (500 MWe), well-controlled power plants will have to be sited at least 50 km from elevated terrain in Class I areas.
- Power plants having capacities of 2,000 MWe or larger, even with SO_2 emission rates as low as 86 ng/J, may have to be sited about 100 km from Class I areas to meet the Class I 24-hour-average SO_2 PSD increment of 5 $\mu g/m^3$.
- In the western United States, which has excellent background visual range, large coal-fired power plants may have to be sited 100 to 150 km from Class I areas to prevent the occurrence of yellow-brown haze in scenic areas during periods of poor atmospheric dispersion.
- Maximum reductions in visual range are estimated to occur more than 200 km downwind of a power plant; therefore, the protection of visual range in Class I areas cannot practically be achieved through constraints on siting alone.

These factors combine to yield regionally variable conclusions about the likely impact of PSD on siting and development. In the West, it appears that Class I areas will provide the limiting constraint in most cases. Latimer, however, concludes that "it may be necessary to limit SO_2 emission from power plants to less than 1 kg/sec, or about 100 tons per day, to meet Class II PSD increments in level terrain in the western United States" (Latimer 1979, p. 64). Latimer also estimates, as previously noted, that Class I constraints and EPA's models can require coal-fired power plants to be as far as 100 to 200 km away from Class I areas.

The larger number of PSD sources in the East combined with a larger ratio of Class II to Class I areas make Class II increments a potentially more limiting constraint there than in the West. At the same time, however, because the background visual range in the East is relatively low, the incremental impact of power plant emissions on visibility will be much smaller in that region.

The range of estimates illustrated in these and other studies underscores the difficulty of making generic rather than case-by-case analyses of the impact of the PSD program. Both the ICF and Latimer reports explicitly recognize this limitation. ICF (1980, p. 1-16) observes:

> the above noted estimated costs are only a preliminary "first-cut" and are not based on site-specific constraints which could increase or decrease the estimated costs. . . . A factual determination of those sources of concern to visibility in specific Class I areas must be based on more detailed, site-specific analyses which are beyond the scope of this report.

Similarly, Latimer (1979, p. 59) emphasizes that because his conclusions:

> are generic in nature, they should not be applied to a given power plant application without consideration of the site-specific emissions conditions, meteorology, and topography.

Assessing the economic costs of PSD even in a specific case may be difficult, since the differential costs associated with BACT and siting decisions may not be readily separable from other costs.

5
THE BACT REQUIREMENT

At any given time the degree of emission control that is economically practical affects the attractiveness of sites for new facilities and the rate at which the PSD increments are consumed. Therefore the state of technology affects the amount of growth that occurs before the increments are exhausted. The more stringent the controls available, the more industrial development can be accommodated in consuming the increments.

The PSD provisions require the use of the best available controls for emissions of each pollutant regulated under the Act. The determination of BACT for SO_2 and TSP is critical to the allocation of SO_2 and TSP increments to a new or expanded source. The Act stipulates that BACT is to be determined on a case-by-case basis during the preconstruction review process, taking into account alternative control strategies, site-specific characteristics, economic effects, energy requirements, and environmental benefits. In no case, however, is the emission limitation permitted to be less than that specified under EPA's new source performance standards (NSPS), the national emission standards for hazardous pollutants (NESHAP), or a State Implementation Plan.

The Act enumerates 28 classes of industrial facilities subject to preconstruction review when such a controlled source emits or has the potential to emit more than 100 tons per year of any regulated air pollutant (see Table 1.3). In addition, any controlled source with the potential to emit 250 tons per year of any regulated pollutant is considered a major source. As of October 1980 NSPS had been promulgated by EPA for 18 of the 28 designated classes of major emitting facilities (Env. Rep. ER 121:1501). In addition, NESHAP have been promulgated for five substances: asbestos, beryllium, mercury, vinyl chloride, and polyvinyl chloride (Env. Rep. ER 121:0461). NSPS and NESHAP are set by the administrator of EPA and include both limitations on emissions and, in certain cases, percentage reductions in emissions from those that would have resulted from fuels not subject to treatment prior to combustion. The NSPS are regarded as being achievable nationwide through the application of the best technological systems of continuous emission reduction, taking into account costs, environmental and health effects not related specifically to air quality, and energy requirements [Sec. 111(a)(1), 42 USC 7411]. Thus, the NSPS, which are reviewed every four years, represent a minimum limitation on emissions--requirements that presumably can be routinely met--although site-specific characteristics of resources may pose problems.

In addition to the general BACT requirement, the Act includes requirements for reduction or elimination of existing man-made visibility impairment in mandatory Class I areas where visibility has been determined to be an important value (Sec. 169A, 42 USC 7491). Existing major stationary sources less than 15 years old that impair visibility in mandatory Class I areas are required to install and operate the best available retrofit technology (BART).

CONTROL TECHNOLOGY

There are a number of concerns with regard to the sufficiency of control technology to meet PSD requirements: whether adequate technology is available for the major emitting facilities that are subject to regulation; costs associated with requirements for this technology; whether the requirements for control provide continuing incentives for development of improved technology; and whether required emission controls are effective for protecting the air quality values of concern.

Availability of technology. What technologies are available for PSD purposes depends in each particular case on a multitude of factors, the most important of which are the state of development of alternative technologies, their costs, and rates of energy consumption. In addition, the adaptability of alternative technologies to specific installations may be affected by characteristics of the site and of the fuels available for use. Under the BACT provisions, the effects of emissions on increments--or on AQRVs in Class I areas--may impose limitations on emissions and thereby determine what levels of performance are required of technology.

Given the current state of technology for controlling SO_2 and TSP, the BACT requirement does not appear to have restrained development of new facilities in the early stages of the program. In fact, permits issued to facilities that burn fossil fuels shortly before revision of the NSPS in September 1979 may have resulted in emission limitations for SO_2 that were too lenient, particularly in those areas of the country with low-sulfur coal. Most of these permits contain determinations of BACT that were equivalent to the NSPS that prevailed when they were written (Dames and Moore 1980), which by 1979 no longer represented the best available control technology.

Because BACT, by its very definition, refers to technologies that are available, the question is not whether technology is available to meet the requirements of PSD but rather how (and when) the capabilities of available technologies will constrain development and restrict otherwise desirable activities in various regions of the country. Regardless of the degree of control, sooner or later there will be a limit to the number of new sources that can be built in a given region without causing a significant deterioration of air quality. So far, when faced with stringent emission limitations, applicants have generally found solutions--technological or otherwise--to satisfy the PSD requirements. There is little question, however, that as increments are consumed, emission limitations will pose major impediments to further resource development.

One option for meeting the BACT requirement that to our knowledge has not been explored is the use of intermittent operational controls. By law, the NSPS are based on continuous controls, but there is no such limitation in the definition of BACT, except that BACT may not be less stringent than NSPS in cases where NSPS exist. Therefore, to the extent that BACT is more stringent than an applicable NSPS, additional intermittent controls would seem appropriate for consideration. Examples of intermittent controls are reductions in operations from normal (or permitted) conditions and fuel switching. The critical factor in defining--and accepting--intermittent controls in a given case is knowing when and how long to apply them. These data will have been obtained from modeling performed for the permit application. Permitting authorities might establish emission limitations that depend on the critical meteorological regimes identified by modeling.

Incorporating intermittent controls in a BACT determination might be attractive where emissions from a facility already using the best available continuous control technology would still threaten to cause increments in a distant Class I area to be exceeded under infrequent meteorological conditions. If the unusual meteorological conditions can be anticipated with acceptable levels of confidence, using intermittent controls in addition to the best continuous controls would protect AQRVs in the Class I area and allow development of the facility that otherwise might be prohibitively costly. Of course, the existence of a facility that otherwise would not have been built--even a well-controlled one--will add to the total regional burden of emissions and hence to regional air quality problems.

Costs. The control technology costs associated with PSD are the additional costs of BACT over and above the costs of control technologies that would otherwise have been required, such as under an NSPS. The additional costs will vary with the type and location of the resources to be utilized. Costs may be expected to increase as increments are consumed, although the BACT provision is intended to provide incentives for development of lower-cost, higher-efficiency alternatives.

All other things being equal, it is reasonable to assume that the BACT requirement implies capital and operating costs higher than those associated with meeting NSPS. Given a specific technology, more stringent limitations in emissions mean higher costs. So far, however, these additional costs have not in fact been great, because the emission limitations of BACT have been essentially the same or only slightly more stringent than those of the NSPS.

The Clean Air Act defines 3 standards of performance for emission control technology on new or modified major stationary sources: The NSPS apply to new sources wherever they are located; the BACT requirement applies in PSD (attainment) areas; and the lowest achievable emission rate (LAER) is required in nonattainment areas. The toughest--and presumably the most costly--limitation is LAER, the purpose of which is to reduce emissions to improve air quality that otherwise is a threat to public health and welfare, as measured by the NAAQS. The least stringent limitations for new sources (according to the definitions contained in the Act) are those of the NSPS, to which all new sources must adhere. EPA has issued guidance documents for determining LAER (U.S EPA 1979d) and maintains a compilation of BACT and LAER determinations (U.S. EPA 1979c). Since BACT determinations are supposed to take energy, environmental, and economic impacts into account and to be at least as stringent as NSPS if applicable, we can expect that BACT will generally be more costly than NSPS but less costly than LAER.

The Act also requires retrofit controls for existing sources in both attainment and nonattainment areas. Under the PSD provisions for protecting visibility, determinations of best available retrofit technology are to take into consideration the cost of compliance, energy and environmental impacts not associated with air quality, existing pollution control technology, the remaining useful life of the source, and the degree of improvement in visibility that may reasonably be anticipated. In nonattainment areas, reasonably available control technology (RACT) must be retrofitted to existing sources, at a minimum, to obtain reasonable progress toward meeting the NAAQS. No general conclusions can be drawn about the relative costs of BART and RACT on the basis of their definitions.

Applicants for PSD permits may also incur costs attributable to PSD other than the additional costs of more efficient controls. One of these costs, cited by many associated with industry, is that incurred as a result of additional delays that may occur in the preconstruction review process. When increments have been or threaten to be consumed, additional costs may be incurred because of the need to locate the proposed facility at an alternative site.

Incentives for improving technology. It appears that the requirement for BACT was included in the PSD provisions of the Act not only to protect air quality but also to provide an incentive for development and adoption of improved control technologies. Has this purpose, in fact, been served? Innovation in control equipment, in response to the BACT requirement, can lead to reductions in the costs of installing and operating control equipment. In most cases, such innovations can be expected to come through progressive increases in knowledge over a number of years of research and development rather than through dramatic technological breakthroughs.

In one case of which we are aware, an applicant, faced with a stringent limitation because of the potential of exceeding an increment, adopted innovative technology which, it turns out, also holds promise of reduced absolute costs. Rather than use conventional SO_2 wet scrubbing techniques, the applicant chose to use a dry scrubbing technique that had only been tested at the pilot stage (Eriksen 1980). Other applicants have subsequently proposed use of this technology too, and thus in this instance the technology-forcing incentives appear to have been successful.

In general, it appears that PSD will serve to encourage development of improved control technologies only as available increments are about to be consumed and perhaps then only after other measures to avoid exceeding the increments (such as relocation or use of alternative models) have been attempted. The driving force for development of new technology will depend on the relative economies of such alternatives. To date, because of the availability of increments, application of the BACT requirement has generally led to modifications of existing control technology to meet site-specific conditions rather than to overall improvement of technological capabilities. Insofar as BACT determinations have in practice required emission limitations equivalent to NSPS (Dames and Moore 1980), the technology-forcing aspects of the PSD program have been diminished. The technology-forcing function of BACT would be more effective if more stringent allocations of increments were a general practice even when consumption of increments is not threatened.

<u>Adequacy of control technology</u>. The BACT requirement imposes emission limitations for all pollutants regulated under the Act. While the BACT requirements are aimed at the criteria pollutants, the effect of the requirement is also to control to a large degree the pollutants that are more directly related to AQRVs. For example, by

controlling emissions of SO_2 and TSP, ambient concentrations of sulfates and other fine particles are also controlled.

The PSD program requires controls on emissions of particulate matter to protect the TSP increments, but fine particles--less than about 2.5 m in equivalent diameter--are of greatest concern for protecting visibility. The BACT requirement applied to TSP is likely to result in improvement of removal efficiency for the larger particles preferentially, since the limitations are specified in terms of total mass and mass varies as the cube of the diameter (for example, removal of a particle twice the size of another results in an eightfold improvement in mass removal). Of course, improving overall mass removal efficiencies is likely to result in improvement in removal efficiencies for the fine particles emitted from stacks. Electrostatic precipitators and fabric filters are routinely used to reduce TSP emissions by more than 99 percent. Studies of removal efficiencies of several of these devices as a function of particle size indicate that they reduce fine particle emissions by 96 to 99 percent (NRC 1980, U.S. EPA 1980b).

BACT is also required for emissions of SO_2 to protect the SO_2 increments and for emissions of NO_x because NO_2, a criteria pollutant, is regulated under the Act. While SO_2 produces some undesired effects as emitted (for example, damage to vegetation), potentially more serious effects are due to the transformation products sulfates, NO_2, and nitrates. Considerable progress in controlling SO_2 has been made over the past few decades both by development of flue gas desulfurization techniques and by using either natural or processed fuels that are low in sulfur content. Lime-limestone wet scrubbers can remove SO_2 with efficiencies of 90 percent or greater (NRC 1980). A dry scrubber has been designed for 90 percent SO_2 removal and has been estimated to cost $23 million less (about half of which is savings in capital costs) over the life of a specific plant than a comparable wet scrubber (Eriksen 1980).

Recent improvements in SO_2 control technology have resulted in revised, more stringent NSPS for SO_2 that include the requirement of a reduction by 90 percent of the potential concentrations of SO_2 in emissions from otherwise high-sulfur fuels (44 Fed. Reg. 33612, June 11, 1979). The requirements for reduction in emissions of

NO_x for coal-fired electric utility boilers depend on the type of boiler and coal and are less stringent than the corresponding requirements for controlling emissions of SO_2, reflecting the fact that available methods for controlling NO_2 are less effective than those for controlling SO_2.

Nitrogen oxides are formed during combustion from both atmospheric nitrogen (N_2) and organic nitrogen present in fuels. Coals mined in the United States contain more fuel-bound nitrogen by weight than distillate oils or natural gas. Substantial improvements in controlling emissions of NO_x formed from atmospheric nitrogen can be obtained by combustion modifications, including staged combustion, and reduction of peak combustion temperatures. Emissions of NO_x from fuel-bound nitrogen, however, are relatively insensitive to combustion temperature, and other combustion modifications, while providing reductions in emissions, have thus far been of limited effectiveness.

Substantial research and development effort is currently directed both at control of the formation of NO_x in combustion processes and at flue-gas treatment for reduction of NO_x emissions; early implementation of both of these control measures is important to PSD objectives.

It is reasonable to expect a significant increase in the use of coal both in generating electricity and in industrial boilers, especially in a number of regions of the country where PSD considerations are a major concern. Owing both to the relatively high nitrogen content of coal and to the relatively less effective controls for NO_x emissions, the increased use of coal is likely to result in a shift in the relative importance of NO_x and its transformation products for air quality compared with SO_2 and its transformation products. Development of improved NO_x controls may well be more important for PSD and other air quality values and a more effective use of available funds than the development of more stringent controls for SO_2 and TSP, since effective measures for controlling SO_2 and TSP are available.

ADMINISTRATION OF BACT

The determination of BACT creates administrative problems for both applicants and permitting authorities.

BACT determinations involve analysis of costs of control alternatives, other economic impacts, energy requirements, environmental consequences of additional controls, and environmental benefits resulting from increased removal efficiencies expressed as reduced use of increments. This balancing of environmental benefits against commercial and energy considerations necessarily involves subjective judgments.

The wide diversity of classes of sources and alternative control technologies that might be considered in BACT determinations requires highly qualified and specialized personnel to analyze the alternatives and decide among them. The Act envisions this process being performed for each permit application, not for classes of sources as a whole. Indeed, site-specific characteristics, the ability of individual firms to pay for PSD controls, and the potential use of incrementally more stringent operational controls pose additional complications and difficulties that cannot be incorporated in BACT determinations unless the analyses are performed for each application.

A major objective of the Clean Air Act is that the administration of its various provisions be undertaken by the permitting agencies in the respective states, to ensure that local considerations are taken into account. Where EPA regional offices administer PSD for a state, decisions are less likely to reflect on local concerns and values. Hence if the objectives of the PSD provisions are to be achieved, the respective states should adopt programs to carry them out locally rather than leaving the job to EPA.

In a number of instances, states and EPA regional offices have established a BACT for a class or classes of sources, ignoring the stricture that BACT is to be determined on a case-by-case basis. We believe it is impractical to expect the EPA regional offices--given both their workload and their understandable reluctance to go beyond minimum requirements of the PSD provisions when acting for a state--to carry out the intent of the BACT process.

EPA currently maintains a clearinghouse of BACT determinations through which interested parties can obtain information on the findings of preconstruction reviews of a wide variety of applications (U.S. EPA 1979c). The usefulness of this clearinghouse depends on the cooperation of state and EPA regional officials, who are its intended users but who also must supply the data and keep it up to date. We urge EPA to continue to expand and develop the BACT clearinghouse and to supplement it where necessary, so that applicants and permitting authorities will be aware of all the options in making case-by-case determinations.

Unfortunately, in many cases state agencies do not have adequate resources of trained personnel and money to do case-by-case BACT reviews. We believe it would be desirable and appropriate for the federal government to provide assistance to the states so that they can carry out their responsibilities under the PSD provisions effectively.

To date there has been a tendency to define BACT controls as no more stringent than those required to meet NSPS, particularly when modeling indicates that emissions will not cause the increments to be exceeded. In the long run this tendency may not be advantageous for the full development of an area's resources. More stringent emission limitations could allow more facilities to be located in a given area for the same amount of degradation in air quality. In cases where there is a threat to the increment, a more stringent emission limitation for BACT--possibly even approaching or equaling LAER--may be desirable, particularly in the light of possible need for subsequent SIP revisions if the increment is found to be exceeded. Under these circumstances, the need for current and consistent information on control technology will become even greater.

NONMAJOR SOURCES

The BACT requirement applies to major stationary sources, new or modified facilities that emit substantial quantities of criteria pollutants. Preventing significant deterioration, however, implies that the measures of importance are the ambient concentrations of certain pollutants and, in Class I areas, the status of AQRVs. Therefore, the Act requires an accounting of the contributions of both major and nonmajor sources to the consumption of increments or to degradation of AQRVs. For this purpose of the Act, maintaining an accurate inventory of nonmajor sources--both point and area-wide sources--is probably of more importance than the precise definition of major sources. The definition of major sources strikes a compromise between comprehensive inclusion of new sources and administrative costs of the PSD program. The exclusion of some sources from preconstruction review does not absolve permitting authorities for accounting for their emissions.

The states and EPA regional offices currently have inaccurate inventories of nonmajor sources. Although many states maintain inventories of all industrial sources, it is not clear how to account for residential and commercial sources. Moreover, there are practical difficulties of applying air modeling to these and mobile sources.

Some states, rather than accounting for nonmajor sources explicitly, reserve a portion of the increments and assume that contributions from the nonmajor sources will not exceed it. This procedure may be inadequate and could ultimately be very costly if modeling at some future date indicates that the increments have been exceeded. If SIP revisions are then necessary, cutbacks in emissions might be required of any or all sources in the area, including PSD-permitted sources already using BACT. Costly retrofit technology or load reductions might then become necessary.

The relative importance of major and nonmajor sources to PSD objectives varies greatly among the different regions of the country, but new nonmajor sources could have an important effect on the status of increments in a significant number of locations, particularly when increments are threatened by emissions from major sources alone. Conversely, replacing old, less well controlled sources with new, well controlled nonmajor sources could result in improvements in air quality. Effective guidelines are needed to help state air quality agencies and the EPA regional offices deal with nonmajor sources consistently and adequately.

POTENTIAL CONSEQUENCES OF BACT PROVISIONS

The Clean Air Act provides incentives that could result in actions inimical to the purposes of PSD. First of all, the LAER provisions of the Clean Air Act, which are applicable in nonattainment areas, by implication require tighter--and hence presumably more expensive--control of emissions than the BACT provisions for PSD areas. Thus the LAER and BACT provisions together create an economic incentive to build new major sources in clean air areas, where less costly controls will be required.

So many factors other than pollution control costs are involved in siting decisions, however, that it is unlikely that this incentive alone will create a substantial problem. Furthermore, in practice there may not be a great difference in costs between BACT and LAER. At least one state has applied a uniform control requirement to both attainment and nonattainment areas to eliminate this incentive [Vermont Statutes Annotated, Title 10, Chapter 23, Subchapter 5-502(3)] and another, California, is considering requiring LAER instead of BACT in PSD areas (California Air Resources Board, undated). The State and Territorial Air Pollution Program Administrators have recommended that BACT be required in both attainment and nonattainment areas (STAPPA 1980).

It has also been suggested that the BACT requirement, along with laxity in accounting for emissions from nonmajor sources in the consumption of increments, could result in incentives to build a number of small sources distributed through a region rather than one large source. Presumably the smaller sources would escape the preconstruction review to which the large ones would be subject.

Again, we are not convinced that operators are likely to find this incentive either attractive or controlling. For the types of major facilities subject to preconstruction review, the diseconomies of small scale are relatively severe. The additional costs of building a number of separate small facilities are likely to be greater in most cases than the incremental costs of BACT relative to NSPS or their equivalents for a single large facility. Furthermore, siting constraints other than those imposed by PSD normally make it impractical to scatter smaller plants at dispersed, less desirable locations. And if the plants were not dispersed, air quality control agencies might regard several small plants under the control of a single operator as one facility, thus making it subject to PSD review in any case.

6
IMPLEMENTATION AND ALTERNATIVES

The objectives of the PSD section of the 1977 Clean Air Act Amendments are to: provide a mechanism for states to make their own judgments about the relative importance of clean air and commercial and industrial development; to protect the national interest in national parks and other federal land areas; and to allocate the scarce resource of clean air where there is no market to make such an allocation.

The PSD program has been in operation for too short a time to pass any overall judgment on how well it is meeting its objectives. Moreover, although PSD is intended to be carried out by the states, only six states have EPA-approved SIPs. Thirteen administer all or part of the federal PSD regulations under delegations of authority from EPA. EPA regional offices administer the federal regulations in the other states and territories. Thus, experience in implementation by the states is too limited to permit conclusions about the extent to which the program is achieving the purpose of encouraging nonfederal decision making.

CRITERIA FOR JUDGING EFFECTIVENESS

Two approaches to institutional strategies for controlling air pollution dominate the professional literature. The first focuses on economic efficiency, the second on administrative convenience. Too often, economic analyses give little or no attention to whether or not the strategy is practical, legal, or administratively acceptable, and administrative approaches, though always pragmatic, are simply descriptive and intuitive, and lack sound guiding principles. In truth, actual strategies for environmental management must include consideration of both administration and economics. Because of the wide variety of situations for which strategies are devised and applied, we would not expect there to be a universally preferred strategy (Crocker 1973). Without suggesting that economic and administrative considerations can in practice be disentwined, for convenience we will examine them separately.

Economic Criteria

The principle on which strictly economic criteria are based is economic efficiency, which, simply put, requires that the objectives of the program be met with the least total cost. An allocation system in which every actual or potential participant acts as if all consequences of his planned activities will accrue to him will be economically efficient. Two implications of the concept of economic efficiency for air polluticn policy are that the incremental costs of pollution control should be set equal to incremental pollution damages and that each additional unit of pollution control should be undertaken by that emitter for whom it is least costly.

The notion of economic efficiency is inevitably tempered by constraints imposed by administrative and technical feasibility. To ensure that each party assumes complete responsibility for the consequences of his acts requires detailed measurement and assessment. But in many cases, we do not know how to trace the consequences of ambient pollution to a particular source or even how to assess the damages done by and the control costs to mitigate this ambient pollution. Even when this knowledge is available, the resources, such as time and manpower, its use would require are not free. Applications of the concept of economic efficiency that disregard the costs of operating an allocation system presume that the benefits of the system are obtainable without cost. A complete perspective requires alternative systems to be evaluated not only on their capacity to cause all parties to bear the consequences of their acts but also on their capacity to economize on the acquisition of information about damages and controls by pollution control authorities.

Since the objective of economic efficiency is unambiguous and analytically coherent, it is convenient to adopt it as a fundamental criterion for evaluating alternative allocation systems; tempering the theoretical construct with such practical considerations as predictability, flexibility, and enforceability (see the section on Administrative and Legal Characteristics).

The economics literature on the efficiency of alternative systems for allocating the scarce air resource focuses on four concerns: cost effectiveness; market distortions, with particular emphasis on entry and exit; incentives to innovate; and decentralization of information.

Cost effectiveness. Analysis of cost effectiveness, which is a limited version of the economic efficiency criterion, asks whether the clean air objective will be obtained at minimum cost in resources that have other valued uses. Attempts to assess cost effectiveness provide insight into the approximate levels at which ambient standards are likely to be efficient.

Economists often criticize technology-based emission standards such as BACT for undermining cost effectiveness. Numerous studies (e.g., Atkinson and Lewis 1976) have demonstrated that the constraints emission standards place on the set of control options from among which emitters may choose have brought about wide discrepencies in marginal costs of control among sources. The greater the differences among sources with respect to size, type, rate of operation and meteorological conditions, and the narrower the legally permissable range of control options, the greater these discrepencies are likely to be.

Where determinations of BACT have in practice taken account of differing circumstances among sources, they apparently have done so on the basis of negotiated determinations of what the applicant can "afford." Sources, such as utilities, that can shift control costs to customers can afford more stringent controls than sources for which the costs must be borne by equity owners. Considerations of affordability thus tend to increase the discrepencies among marginal control costs for different sources and thereby work against cost effectiveness.

Market distortion. Market distortion may be caused by an allocation system for the air resource. For example, an allocation system that dictates pollution control equipment without regard to the economic consequences for the emitters may inadvertantly reduce competition in the sales markets of the emitters. Existing firms that lack ready access to sources of financing may have to withdraw from the market, and potential entrants may be discouraged from trying to compete with firms already solidly entrenched. Allocation systems can conceivably be constructed so that they grant or further monopolistic and monopsonistic advantages, preserving some competitors rather than competition and violating the notion of economic efficiency.

Incentives to innovate. The private market often fails to provide rewards for innovation in pollution control, and some allocation systems for pollution control intensify this tendency. The standard example is an allocation system defined only in terms of emission and technology-based standards: once an emitter has met a standard, he gains no reward from further emission reductions. At best, any innovative effort he does make is likely to be biased toward that which reduces the cost of reaching the predetermined emission limitation. Control innovations that reduce emissions without affecting control costs will be neglected.

Indeed, if he is successful in a search for an emission-reducing control technology, and if the emissions standard is defined in terms of some concept of the "best available technology," he pays a penalty: his allowable emissions will also be reduced. In effect, the responsibility for discovering and evaluating new technologies in allocation schemes based on technology-based standards such as BACT then rests with the pollution control agency. Repetto (1980), however, demonstrates that the existence of a distinct, competitive industry that develops and markets pollution abatement equipment will respond with innovation under a system of technology-based emission limitations, since this industry has incentives to reduce costs of emission control for its customers.

Decentralization of information. Decentralization of information can be loosely defined as a situation in which knowledge about one's own state of affairs is not directly communicated to any other party. Communication about the benefits and costs attached to alternative acts occurs through participation in a process (such as a market). Participation in this process collapses all one's benefits and costs into one or a very few scaler values (such as a bid or ask price). The amount of detailed information that any single party must acquire to participate in a market is considerably less than the amount a centralized agency would need if it were to try to substitute for and duplicate the results of a market. Furthermore, the information embodied in the communicated signal (such as price) is more refined, given that one is more familiar with the peculiarities of one's own circumstances than is anyone else.

And, although it is not immediately relevant to economic efficiency, it is worth noting that no single party alone controls all the information an economic system produces. Monopoly over information can constitute power. At most, the function of the central agency in a system based on economic incentives is limited to strategic intervention in the allocation process; the agency cannot try to set and actually steer the course to be followed. At a minimum, the agency must initially put together the allocation process. If the agency is unburdened with the details of supervision, it can assume responsibility just for designing incentives that are apt to guide emitters and receptors more effectively. Otherwise, the presumed benefits of greater autonomy will not be captured.

Administrative and Legal Characteristics

Several administrative and legal characteristics will help to determine whether the PSD program is sound public policy and whether it can be successfully implemented. Chief among these are clarity, predictability, flexibility, enforceability, and degree of support.

Clarity. Ideally, the goals of a public policy should be clear and the relationship between implementing mechanisms and goals should be readily understandable. If these criteria are not met, the support and cooperation of the many individuals and groups in and out of government necessary to make a program viable is likely to be weak or nonexistent.

The seemingly basic characteristic of clarity may be the Achilles' heel of the PSD program. The goals of the program are not clearly delineated in the statute and, insofar as the law tries to define the program's goals (for example, protecting AQRVs), the inadequacy of scientific knowledge makes it impossible to match the goals with a clear set of program requirements. The result, perhaps inevitable given scientific uncertainty and the great diversity of existing conditions and public values, is the merging of means and ends. The objective of the program is to establish a process rather than to mandate a specific air quality goal.

The concept of PSD as a process is subtle and to many people elusive. Not only industry spokesmen but a number of state officials told us that there was no need for PSD--the secondary ambient air quality standards were sufficient to protect against adverse effects from air pollution. This interpretation implies that PSD will be resisted by both the industries that must comply with its provisions and by some of the government officials charged with carrying out the program.

Predictability. The complexity of PSD means that some aspects of the program are not very predictable. A lack of predictability discourages voluntary compliance (which is necessary for the success of any regulatory program) and tends to detract from the respect with which the program is viewed. It may also result in inequities, with some pollution sources being arbitrarily subjected to different marginal costs than others. The three main points of unpredictability in the PSD program are air quality modeling, the BACT review, and applying the AQRV test in Class I areas.

The difficulty of verifying models by monitoring, the scientific crudeness of most of the models used, and the lack of certainty surrounding the meteorological and other data used with the models makes the reliability of any particular modeling effort uncertain (see Chapter 3). Thus, for example, when an initial modeling effort concludes that increments have been consumed, it seems reasonable to expect that an applicant will try a different model. Modeling is a rapidly developing art, and there are many models from which to choose. The models recommended by EPA in its guidelines are limited in applicability and, by virtue of the time it takes to prepare and establish formal guidelines, are likely always to lag behind the state of the art. Furthermore, as described in Chapter 3, there are inherent problems with demonstrating comparability of models when the standard model has deficiencies that the newer model is intended to overcome. Modeling, instead of providing the factual context for policy decisions, can thus be used as a means for justifying a desired outcome.

The BACT review is supposed to be individualized for major new sources (see Chapter 5). BACT review is intended to be a negotiating process, and if the outcome of a negotiation were totally predictable, no negotiation would be necessary. In practice, however, this aspect of PSD seems to be more predictable than one might expect, because of the tendency of state and federal officials to define BACT for categories of sources and often to equate BACT with an NSPS where an NSPS exists.

Greater uncertainty surrounds the issue of how federal officials will discharge their responsibilities to protect AQRVs in Class I areas, because except for visibility, which is fairly well understood, there are currently uncertainties about the relationships between atmospheric concentrations of pollutants and perceptions of AQRVs (see Chapter 2).

There are aspects of PSD that are and should be predictable. At least since Alabama v. Costle, it is reasonably clear which sources are subject to preconstruction review. It is also clear which areas are classified as Class I, although the question of whether a new or modified source will adversely affect a Class I area, and to what degree, can be answered only through modeling.

Flexibility. Flexibility, which has its own virtues, is the other side of the coin of predictability. A program that is too rigid becomes incapable of dealing with the variety of circumstances that are likely to arise in a large and heterogeneous advanced industrial society.

If lack of clarity is the most salient weakness of PSD, flexibility is probably its strong point. The problem faced by architects of the program was how to devise a system to protect values on which there was no national political consensus (other than that the values should be protected) and about which there was a great lack of scientific knowledge. Furthermore, the program had to operate in a multitude of air quality control regions, which differed greatly in the extent and nature of existing pollution and which would be subjected to a wide variety of development pressures. The solution was to develop a process that gives the states considerable flexibility.

The greatest flexibility is in the BACT review of individual sources. Not only can technological requirements be matched to individual plants, but even the type of major source subject to review can to some extent be tailored to local conditions. This flexibility in BACT determinations is a source of uncertainty in the requirements that applicants must meet to obtain permits.

The states also have some flexibility to change the classification of areas. How much flexibility is not yet known, because so far as we are aware there has been only one reclassification from Class II to Class I (Northern Cheyenne Indian Reservation in Montana) and there is pending only one reclassification from Class II to Class III in three counties in southwestern Illinois. The potential exists for other reclassifications as well. For example, the Secretary of the Interior has determined that AQRVs are important attributes in 44 areas of federal land that are now designated Class II (Figure 6.1); the implication of this finding is that state authorities should consider these areas for reclassification (45 Fed. Reg. p. 43002, June 25, 1980).

Flexibility is also implied by the waiver and variance procedures with respect to Class I increments and by the reclassification procedure. Again, two desirable features, predictability and flexibility, are in conflict. Predictability simplifies the business decision but flexibility may ultimately lead to more opportunities for industrial development.

Enforceability. It is reasonable to assume that the emission limitations of PSD permits can be enforced, just as other types of emission limitations are enforced. The question remains whether permitting authorities--either the states or the EPA regional offices--will be able to keep track of the consumption of increments or to remedy violations. Keeping track of the status of increments involves maintaining an accurate inventory of both major and nonmajor sources and modeling the contributions of all sources to degradation of air quality. Neither of these tasks is technically or administratively trivial. More difficult still will be the task of reducing emissions if it is found that increments have been improperly exceeded for one reason or another. Under such circumstances, a permitting authority would be required to determine which sources were responsible and to develop and implement plans to remedy the situation. Such a plan could include rollbacks--retroactive

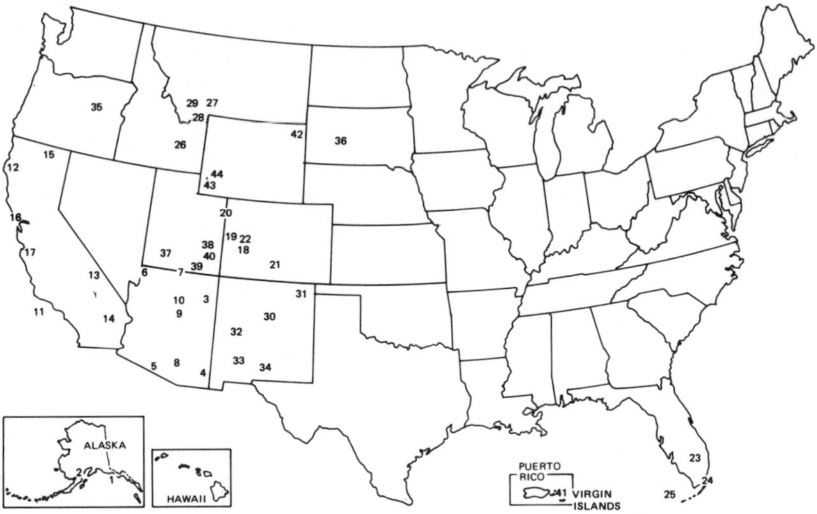

Alaska
- Glacier Bay NM 1
- Katmai NM 2

Arizona
- Canyon de Chelly NM 3
- Chiricahau NM 4
- Organ Pipe Cactus NM 5*
- Paiute PA 6
- Paria Canyon PA 7
- Saguaro NM 8*
- Sunset Crater NM 9
- Wupatki NM 10

California
- Channel Islands NM 11
- Chemise Mountain PA 12
- Death Valley NM 13
- Joshua Tree NM 14*
- Lava Beds NM 15*
- Muir Woods NM 16
- Pinnacles NM 17*

Colorado
- Black Canyon of the Gunnison NM 18*
- Colorado NM 19
- Dinosaur NM 20
- Great Sand Dunes NM 21*
- Powder Horn PA 22

Florida
- Big Cypress NPr 23
- Biscayne NM 24
- Fort Jefferson NM 25

Idaho
- Craters of the Moon NM 26

Montana
- Beartrap Canyon PA 27
- Centennial PA 28
- Humbug Spires PA 29

New Mexico
- Bandelier NM 30*
- Capulin Mountain NM 31
- El Morro NM 32
- Gila Cliff Dwellings NM 33
- White Sands NM 34

Oregon
- John Day Fossil Beds NM 35

South Dakota
- Badlands NM 36*

Utah
- Cedar Breaks NM 37
- Dark Canyon PA 38
- Grand Gulch PA 39
- Natural Bridges NM 40

Virgin Islands
- Buck Island Reef NM 41

Wyoming
- Devil's Tower NM 42
- Fossil Butte NM 43
- Scab Creek PA 44

*Part of this area is already designated Class I.
NM = National Monument
NPr = National Preserve
PA = Primitive Area

FIGURE 6.1. Areas recommended for reclassification to Class I by the Secretary of the Interior.

reductions in emission limitations--that might be difficult to adopt and enforce.

The most critical enforcement problem will arise when an increment for an area has been completely consumed and there are additional major facilities that want to locate in the area. This seems to be the situation now in western North Dakota, and it will occur in an increasing number of areas over the next five or ten years. Whether state air quality agencies or EPA can actually prohibit any major new development will depend on how much political support exists for the PSD program.

Degree of support. The degree of support enjoyed by a program is a crucial variable in determining its success. It is a reflection of the other criteria, because a program that is successful in other ways is likely to enjoy support.

Congressional support for PSD is currently mixed. Among interest groups, the environmental community tends to support the program strongly while industry tends to oppose it. The attitudes of state air quality officials are quite diverse, ranging from reluctance to accept the program through businesslike implementation to wholehearted embrace and innovation. The situation, in short, is volatile, and much will depend on how the program is implemented in the coming months and years.

ALTERNATIVES TO THE EXISTING SYSTEM

In evaluating a program such as PSD, it is useful to consider alternatives to the existing system, both to assess their potential advantages and to gain a better understanding of the merits and flaws of the current system. The economics literature divides allocation systems into two somewhat overlapping classes: regulatory or command systems and economic incentive systems. Although analyses of these alternatives usually assume the two are mutually exclusive, combinations of the two deserve attention.

The fundamental difference between the two kinds of systems resides in the manner in which controls are exercised. Pure regulatory systems establish rules that specify quantities such as emission rates. Violations of the rules are defined in terms of deviations from these predetermined quantities. Noncompliance engenders judicially imposed injunctions and/or financial penalties. Economic incentive systems, on the other hand, depend upon some use of prices to influence quantities.

To operate economically, both systems must have essentially the same information on control costs, damages, and responses of emitters and receptors to altered circumstances. Most of the relevant information initially rests with emitters and receptors (or their agents). In some economic incentive systems, these parties retain it and use it to establish whatever responses to price variations they think are most fitting. In a purely regulatory system, the information held by emitters and receptors is somehow acquired by the control agency. The agency uses this information to perform the calculations that would otherwise result from the interactions of participants in a market. Since the agency presumably has fewer resources with which to perform these calculations than the aggregate market participants, less information is brought to bear on decisions in purely regulatory allocation schemes.

At present, however, knowledge about what, if anything, would in practice be gained by moving from a regulatory to an economic incentive system for controlling air pollution is nonexistent for PSD applications and sparse for nonattainment areas. Studies of the relative advantages of economic incentives over regulatory schemes for controlling air pollution have dealt exclusively with nonattainment areas, where typically there are many sources of different types; the applicability of the results of these studies to PSD areas is therefore limited. In a study of St. Louis, Atkinson and Lewis (1976) found that different versions of perfectly functioning economic incentive systems would be three to ten times as cost-effective in achieving the primary NAAQS for particulates as is the currently enforced state implementation plan. Anderson et al. (1979), in a study of a proposed short-term NO_2 standard in Chicago, obtained nearly identical results. Nevertheless, for some types of economic incentive systems (e.g., subsidies or the much-discussed emission charges) it is difficult to discern how anything would be gained except perhaps increased incentives to innovate.

To be economically efficient, effluent charges or control subsidies require no less information than a set of emission standards tailored for each source (Rose-Ackerman 1973, Marin 1978). It is sometimes suggested that an agency could avoid much of the information problem for emission charges by initially establishing a charge level and then manipulating it until the desired ambient pollutant concentrations or total system emissions are attained. The dynamics and stability properties of this adjustment process, however, are ill-understood. If the charge an emitter will face in the future is dependent in some known way on his present behavior, he will have an incentive to invent various forms of nonproductive strategic behavior. Moreover, the uncertainties emitters would face as an agency tried to discover the "optimal" price could be very costly, especially given the heavy fixed investments required for most control technologies. Emitters would bear the costs associated with erroneous decisions made by the control agency.

In recent years, economists have attempted to identify circumstances where, despite desirable incentives for innovation, the efficiency gains from economic incentive systems are problematic. Included are circumstances where market distortions exist. The emissions tax, for example, can cause an emitter who also has a monopolistic advantage in his salable output to reduce the value of his total output by more than the value of prevented pollution damages (Baumol and Oates 1975). If it is technically impossible or extremely costly to monitor emissions, one must resort to direct controls on (or the pricing of) inputs, production processes, tail-end equipment, or source location. Another problematic circumstance is when incremental damages change very rapidly--for example, when small variations in pollution are quite harmful (Yohe 1977). Emission taxes are likely to allow the emitter more freedom to vary his emissions and could thus allow large damage effects with even minor shifts in economic influences on the level of emissions. Chapter 2 presents some limited empirical evidence that air pollution effects upon atmospheric visibility and aquatic ecosystems at low ambient pollution concentrations are in fact characterized by rapidly changing marginal damages. Conversely, if marginal costs of control climb very rapidly, with marginal control benefits being more-or-less constant, the emissions charge becomes economically appealing. With the charge, emission levels can vary with changes in economic parameters, thus preventing control costs from rising to disproportionately high levels.

Public utilities are another troublesome case. They are guaranteed some "fair rate of return" over their costs, so to the extent that they can pass the emissions tax on to their customers, incentives to reduce emissions are reduced. Much would depend, of course, on determination by the rate-setting authority of what constitutes a cost.

There is an economic incentive strategy that has many of the desirable properties of the direct regulatory strategies (Crocker 1966, Dales 1968). In this strategy, generally known as the salable emission permits strategy, the control agency issues permits considered compatible with the predetermined level of ambient air quality. These permits could then be traded, just as the title to a piece of land is traded. The prices that emerged would embody relevant information about the peculiar circumstances of each emitter and would automatically adjust to altered economic conditions. The functions of the control agency would be to organize the market, to ensure that the market is and remains reasonably competitive, and to maintain the integrity of the properties represented by the emission permits.

Under a marketable permit system, the burden on the control agency of gathering information is reduced from that under a regulatory or emission charge system. Changes in available technologies and economic conditions would be registered in permit prices. Thus the parties involved in the allocation of the air resource would respond to changes by altering their behavior in an economically efficient fashion, assuming there were no persistent, systematic biases in the market's predictions of the course of future events. There would be a continuing incentive to reduce emission control costs and thus to innovate in control technology, because improved, lower-cost control technology would reduce the number of emission permits to be purchased. Finally, Anderson and his colleagues (1979) have presented limited empirical evidence that the legal and technical machinery required to administer a salable permit program would be similar under a regulatory system or an emission charge system, and the number of emitters requiring control would be fewer. If true, these two features should enhance the political appeal of the salable permit system.

The proponents of a salable permit system have been vague, however, about important details of the system. Few efforts have been made to provide bases for deciding the geographical area or temporal span to which a permit applies; who, when, and how an interested party can participate in the market; how the agency will dispense of permits; the role the agency might play in "making" a market when there are few participants; the structure of institutions to span permit sales and purchases across time; under what conditions, if any, the agency should play the role of a trust-buster when an emitter gains monopolistic advantages by cornering the emissions permit market of an area; the penalties, if any, to be imposed upon nonholders of permits who emit; how the stock of permits would be adjusted; and other nontrivial matters.

Good analyses (e.g., Anderson et al. 1979) of many of these issues have been carried out for proposed salable permit systems in nonattainment areas having large numbers of existing sources. PSD areas, however, are characterized by small numbers of sources and infrequent requests for permits to add to existing emissions. These new sources, when they do make their infrequent appearances, tend to have the potential to consume very substantial portions of the allowable increment. Thus, it is by no means obvious that in PSD areas the economically desirable properties of competition will emerge through either the initial distribution of permits by auction or to the exchange of permits in a market.

If some emitters (such as nonmajor sources) were exempt from the requirement of obtaining permits, the number of potential buyers and sellers would be further limited and the market would thus be even less competitive. Transactions costs would be increased, because the search for potential trading partners would be more difficult, and middlemen who would "make" markets for permits would be less likely to emerge. Monopolistic advantages would more easily accrue to large sources.

The greater the precision with which the scope of a permit is legally defined, the lower the transactions costs and thus the greater the likelihood that a system of salable emission permits would simultaneously allow more industrial growth, net improvements in air quality, and economically efficient outcomes. All legal and administrative requirements that make transactions cumbersome--thereby increasing transactions costs--make it less likely that a salable emissions permit system will succeed. If salable permits were related to effects on air quality, the complexity of modeling would make the cost of exchange quite high, because of the difficulty and delay in determining the effect of the permit exchange on air quality.

If transactions costs and monopolistic advantages are not relevant, economic efficiency is not affected by who initially gets emission permits. The market will ensure that permits ultimately reside with those who value them most highly. But both transactions costs and monopolistic advantages reduce the set of mutually beneficial exchanges and decrease the availability of permits to those who do not hold them; if these factors are substantial, attaining economic efficiency requires an efficient initial allocation of permits.

The large scale and small numbers of new sources in PSD areas are not likely to foster competitive auctions of salable permits. In these circumstances, it seems likely that the control agency would intervene to restrict the number of permits or their denominations that are available in the initial auction. In effect, this intervention determines the degree of control for the applicant, as the preconstruction review does under the current system.

If the agency wishes to maximize both AQRVs and the value of industrial activity, it is faced with a severe information problem. In addition to AQRV effects, the agency must know the applicant's rules of behavior under uncertainty, his current and expected least-cost control technologies, the input and salable output markets to which he has access, and, finally, the manner in which the expected activities of future applicants, if any, will influence these factors. Practical problems associated with implementing a salable emission permit system for PSD purposes therefore lead to the need for air quality control agencies to perform the same types of analyses and impose the same types of contraints on emissions as they must for preconstruction review under the current PSD system. Thus it seems unlikely that a salable permit system would be advantageous compared with the current regulatory system in attainment areas.

POSSIBLE CHANGES IN THE EXISTING PSD PROGRAM

Short of replacing the existing PSD program with a radically different one, there are numerous ways in which the existing system could be changed or modified. That is to say, the regulations, and to some extent the provisions of the law itself, could be altered to fulfill the purposes of the law more effectively.

Pollutants Covered

All of the pollutants--both Set I and Set II--encompassed by the PSD program have some relationship to AQRVs and are appropriately included in new source reviews under PSD.

Two categories of pollutants, however, are not analyzed in the program: fine particulate matter and sulfates (see Chapter 2), both of which are major contributors to visibility impairment, acid rain, and other important air pollution problems. To some extent, consideration of sulfates can be subsumed under considerations of fine particles; therefore we will not discuss sulfates separately.

Fine particles, defined as particles smaller than 2.5 μm in diameter, are of major concern in the context of PSD and for the entire air pollution control effort (Perera and Ahmed 1978, NRC 1979d). Fine particles are a major element in visibility impairment as well as in other adverse effects of pollution. Total suspended particulate mass is relevant to the visibility question only insofar as TSP regulations cover fine particles. In most cases emission limitations for TSP will result in elimination of a major portion of the emissions of fine particles, although the control mechanisms are designed to maximize the reduction of emissions of TSP mass, not numbers or mass of fine particles (NRC 1980). To ensure a more direct relationship between the PSD preconstruction review and evaluation of AQRVs, especially visibility, we recommend the establishment of an increment for fine particles themselves (see Chapter 2).

Major prerequisites to implementation of this new increment would be the development of air quality models to take account of the formation of fine aerosols in the atmosphere from precursor gases (Henderson et al. 1980) and the characterization of fine particle emissions from major and nonmajor sources (NRC 1980). Making explicit the information most relevant to the more important AQRVs would place permitting authorities and federal officials in a better position to make the difficult and complex trade-offs between economic and environmental values.

Baseline Date

The decision in Alabama Power v. Costle decreed that in the federal program the date for which the baseline concentrations are to be assessed for each PSD area shall be the date when the first permit application for that region is filed. Thus, the baseline date is likely to be different in each PSD area. It probably would have been preferable, from the standpoint of administrative convenience and simplicity, to have a uniform nationwide baseline date as originally proposed by EPA.

Additional confusion resulting from the Alabama Power decision relates to a change in definition of major source. If a PSD permit had been filed in an attainment area for a source that was a major one under the old definition but not a major one under the new definition, was the baseline date triggered by that application? If so, then baseline dates will have been established for many, if not most, PSD areas, and there will be little confusion. If not, then a new baseline date will have to be determined in many attainment areas.

Increment Levels

Would a change in the levels at which the TSP and SO_2 increments are set improve the PSD program? This is a difficult question, because the increments contained in the statute are somewhat arbitrary--that is, they have only a limited relationship to AQRVs. Any alternative set of nationally uniform increments would be equally arbitrary, because adverse consequences for air quality can only be related to ambient concentrations, which depend on both the baseline concentrations and the extent to which the increments are consumed. Furthermore, the scientific knowledge to relate low ambient concentrations unambiguously to adverse effects is lacking except in the case of degradation of visibility.

The question is further complicated by different conceptions of the functions served by the increments. On the one hand, they can be viewed as mechanisms to trigger consideration of area classification, as a way of allowing an area to make an explicit decision about how clean the state wants its air to be. On the other hand, if reclassification proves to be difficult or impossible, then the increments serve as absolute limits on air quality deterioration and greatly increase the costs of industrial development, possibly making it economically unattractive.

A system of increments combined with a case-by-case review of emission limitations seems to be as effective a way as any to maintain air quality at levels better than the NAAQS in order to protect national values (as in the case of mandatory Class I areas) and to implement local values in areas that desire cleaner air than minimally required. Since any other specific increments would serve the same purpose as the current ones, we cannot recommend any changes in the current numbers.

Increment Averaging Times

The statute now includes increments for annual and 24-hour averaging times for TSP and annual, 24-hour, and 3-hour averaging times for SO_2. The averaging times are as important for the stringency of the increments as the actual levels at which the increments are set. At least for SO_2 and TSP, the 24-hour increments are the most constraining on most new sources that affect Class I areas; of Class II increments in the vicinity of a source, the 3-hour SO_2 increment is the controlling one.

As described in Chapter 3, establishing just one averaging time--that is, one increment--for each pollutant would greatly simplify the PSD process. The modeling for a new source would be easier, and depending on the averaging time chosen and modeling capabilities, the impact of PSD could be more predictable for both regulators and the regulated. As argued in Chapter 3, a single averaging time could provide the same protection as a set of averaging times if the levels of the increments were consistent with the same distribution of air quality.

Because the occurrence of a short-term maximum concentration is a rare event, the results of alternative models for short-term maxima may differ considerably while the results of different models for long-term averages usually agree rather closely. The use of long-term averaging times (annual or seasonal) in the increments would, therefore, have the likely effect of reducing the differences among models.

We do not have sufficient information to recommend a specific averaging time, but we do recommend that the existing averaging times be reviewed with the goal of reducing the number of increments. Reducing the number of increments need not decrease the stringency of PSD requirements.

Statistical Basis of Compliance

In Chapter 3, we argued that it appears desirable to express the criterion for compliance with short-term air quality standards and increments in terms of the probability of exceeding a given concentration rather than as a limit that can only be exceeded once. The statistical reliability of the results of air quality modeling would be greatly enhanced by such a change, which can be accomplished without altering the levels of protection afforded by the current numerical values of the increments and NAAQS. (A change in the compliance criterion for the increments would be unnecessary, however, if all the short-term increments were deleted in favor of long-term increments.) Because the consequences of such a change on all air quality programs would be far-reaching, we recommend further study of this proposal.

Sources Subject to Review

The definitions of the major sources that will be subject to PSD review appear to us to be sensible and appropriate. However, the definition of major sources and the requirement that only major sources be subjected to preconstruction review raises concerns that the impact of nonmajor sources, including mobile sources, on air quality in PSD areas may not be paid adequate attention. Simply put, the cumulative effect of a number of minor sources locating in an area could well be great enough to cause significant deterioration of air quality.

Applicants for PSD permits should therefore be required to include emissions from minor sources in the models used to support their applications. Some existing PSD programs do this; all should. To make this possible, most states would have to do a better job of maintaining inventories of minor sources. Such inventories of new minor sources should be included in state implementation plans.

Mobile sources present a difficult problem because they clearly cannot be subjected to the permitting process used to deal with major stationary sources. EPA is now wrestling with this problem in trying to establish regulations for the so-called Set II pollutants, most of which come largely from mobile sources. At the very least, estimates of the emissions resulting from changes that affect mobile sources (such as new highways) should be included in the models used to support PSD applications, just as nonmajor stationary sources should be included.

A requirement that analysis of the effects of emissions from nonmajor stationary and mobile sources be included in all PSD permit applications would result in a smaller portion of the increments being available for major sources. This requirement would mean major sources might have to install somewhat more stringent controls or, at worst, there could be some reduction in the number or size of major sources built in PSD areas. The distinction between major and nonmajor sources, however, is largely a matter of administrative convenience and a realistic recognition of constraints on the resources of the permitting authorities. These considerations should not be allowed to subvert the basic intent of PSD, which is to regulate emissions that may cause significant deterioration, regardless of the size or type of source.

Modeling and Monitoring Requirements

There is no way to estimate the effect on air quality that a proposed new source will have without using air quality models. In some cases inappropriate models have been used under the PSD program, but we believe that the basic modeling requirements in the statute are sound and should not be changed at this time.

The monitoring requirements are a different story. Section 165(e)(2) of the Act requires either the state or the permit applicant to gather continuous air quality monitoring data for a period of a year prior to application for a permit. The purpose of such monitoring, according to the Act, is to determine whether emissions from the facility for which a permit is sought will exceed either the increments or the NAAQS. But in fact, such determinations will be made by modeling and not by monitoring. We therefore fail to see how general preconstruction monitoring of ambient air quality serves a useful purpose for analyzing potential effects of proposed sources.

Two other suggested purposes for preconstruction monitoring are to provide the data for determination of the baseline and to provide "a certain discipline on the use of modeling techniques. . . [through] a continual process of conformation and reassessment, a process that enhances confidence in modeling, as a means for realistic projection of air quality" [Alabama Power Company v. Costle, 13ERC 1993, 2019 (D.C. Cir. 1980); see also id. at 2022]. The first purpose applies only to the initial applicant for a permit. The second purpose, while important, may just as well and more easily be served by postconstruction monitoring requirements. Such monitoring, as well as monitoring to determine if maximum allowable concentrations (i.e., the NAAQS) have been exceeded, should presumably take place as part of general air quality monitoring programs. Furthermore, preconstruction monitoring to validate models, while an important and useful activity in some cases, is not likely to be necessary in every case. Amendment of the statute to provide flexibility to permitting authorities in imposing preconstruction monitoring requirements would make the permitting process simpler, faster, and less expensive without reducing the effectiveness of PSD.

The EPA PSD regulations also require postconstruction monitoring, at least of TSP and SO_2. There may be situations where such monitoring can be used to check the validity of some of the modeling, but these opportunities are very limited, owing to the nature of the assumptions that must be employed in modeling. Thus any blanket requirement for postconstruction monitoring should also be carefully evaluated.

Preconstruction collection of meteorological data (including wind speed and direction), on the other hand, would be highly useful for analyzing the potential impact of a proposed new source. Collection of this type of data should be encouraged or required.

Emission Limitations

The BACT provision of the Act requires case-by-case determination of the emission limitations to be met by applicants for PSD permits. Case-by-case assessments of control requirements offers opportunities for balancing environmental and developmental values at the state level and for accommodating special needs; it also introduces uncertainty into the process of acquiring a permit (Arthur D. Little 1980).

If our interpretation of PSD as a process is correct, the flexibility of case-by-case determinations of BACT is worth the increased uncertainty. Incorporating intermittent controls in BACT determinations would be practical only on a case-by-case basis. Case-by-case determinations also provide more of an incentive for the development of innovative technology than would uniform BACT determinations. We therefore regard the current BACT provisions as appropriate and necessary to achieve the objectives of the Act.

Area Redesignation

If our view of the PSD program as a process rather than a set of rigid standards is correct, it follows that there should be a good deal of flexibility for a state to change the designation of an area. There have been proposals for such redesignations. The Secretary of the Interior has identified 44 areas where the importance of AQRVs suggests consideration of redesignation from Class II to Class I (Figure 6.1). The California Air Resources Board has held public meetings to discuss redesignation of 30 places now in Class II areas to Class I. Illinois is considering a proposal to redesignate 3 counties in southwestern Illinois near St. Louis from Class II to Class III (Illinois EPA 1980). It is too early in the program, however, to tell whether there will in fact be flexibility in changing designations. If it proves impractical to reclassify lands, then amendments to the law and the EPA regulations to ensure that reclassification is a practical tool should be considered.

Waivers and Variances

As in the case of area redesignation, we believe that flexibility is intended--and desirable--in the procedures for and decision making on the waivers and variances provided by the Act. Here there is no experience from which to judge whether such flexibility will actually exist. There are two points of particular concern.

First, section 165(d) of the Act specifies that the federal land manager can recommend that the state not issue a permit for a new source, even if the projected emissions would not result in an allowable Class I increment being exceeded, as long as he is convinced that AQRVs would be adversely affected. On the other hand, the Act allows the federal land manager to determine that AQRVs are not threatened even if the increment is exceeded; in this case the permitting authority may grant a waiver and issue a permit. To make informed decisions the federal land managers must appreciate available scientific knowledge about the effects of changes in air quality on AQRVs and must have guidance on how to make these decisions. The general community concerned with PSD and its implementation needs to know how federal officials will assess AQRVs so actions on requests for variances can be anticipated or understood. We therefore recommend that the responsible federal agencies develop and publish guidelines that specify how they intend to assess the importance of AQRVs and under what conditions the respective federal land managers will recommend permits be denied or granted.

Our second concern has to do with sources that may result in concentrations that exceed the increments temporarily. The Act recognizes and allows temporary emissions associated with the construction phase of a major project. The national interest is likely to be served in the long run if emissions from relatively short-lived projects are treated similarly. A small oil and gas field, for example, produces most of its emissions over a two or three year period and emission rates fall as the field is depleted. In keeping with the intent of the Act, such sources should be treated on a case-by-case basis.

Retrofitting

Section 169A(b)(2)(A) of the Act requires use of the best available retrofit technology on existing sources that adversely affect visibility in certain mandatory Class I areas. This requirement is a realistic recognition by the Congress that visibility in some areas of national importance is impaired now, and that control of new sources alone would not be sufficient to deal with the problem.

The same reasoning that led to the BART requirement for visibility protection is applicable to the acid rain problem. There is good evidence that acid rain is a serious problem now and that it has been getting worse over the past couple of decades (Likens et al. 1979). The current acidity of precipitation in large areas of the country cannot be managed by controlling only new sources unless we are willing to wait thirty or forty years to alleviate the problem. The major sources that contribute to acid rain, like the sources that impair visibility, are large, capital-intensive facilities (power plants and smelters, for example) with long lifetimes. Controls imposed on new sources will extend the lifetime of existing facilities even longer, because such controls provide an economic disincentive to construct new plants. A retrofitting requirement would remove this disincentive in most cases.

We recognize that retrofitting existing plants is sometimes an expensive and difficult undertaking. Whether the expense is worth the benefit of dealing now with visibility, acid rain, and perhaps protecting other AQRVs is a political and social decision. There should, however, be no illusion that controlling only new sources will deal adequately with these problems any time in this century.

There is another major difficulty in dealing with visibility and acid rain. Both problems involve the transport of pollutants over hundreds, sometimes thousands of miles. The Clean Air Act is structured to deal with pollution sources whose adverse effects occur in the immediate vicinity of the source. But the acid precipitation in the Adirondacks may in large part be due to sources in Ohio, and significant contributions to visibility impairment in Maine may be produced by sources in the vicinity of New York City. Some basic rethinking of the assumptions underlying the Act will be necessary if our air pollution control programs are to deal with such long-range transport phenomena.

CONCLUSION

The discussion of possible changes in the PSD program underlines the importance of how PSD is interpreted. Our impression is that both industry and regulatory agencies have interpreted the PSD provisions as a set of restrictive and rigid federal standards. This is not our view of PSD.

The program does provide relatively inflexible federal protection for areas of national importance, such as major national parks. But as a result of our study, we have concluded that for other areas that are meeting the national air quality standards, PSD is intended to be a process that provides states with the flexibility to implement their own diverse views about the value of clean air.

In keeping with this interpretation, we would encourage those states that have not yet done so to obtain PSD permitting authority from EPA. Once the states have such authority, decisions on PSD permits and variances are more likely to reflect local conditions. This may, of course, mean more or less stringent protection of existing good air quality, depending on the outcomes of the political processes in the respective states. Permitting authorities should be innovative in using the flexibility that we believe is inherent in the current system.

It is possible that the rigid interpretations of PSD will be self-fulfilling. If the flexibility that we think is intended to characterize the program does not materialize then more radical changes than those we have proposed might have to be considered. But to date there has not been sufficient experience with the program to know whether that flexibility will materialize.

The inadequacy of current scientific knowledge is a major reason for the complexity of PSD. New knowledge needs to be developed and existing information needs to be consolidated and evaluated. To use the current PSD provisions to maximum advantage, applicants, permitting authorities, and federal land managers need guidance on how to assess the importance of air quality related values and the effects of changes in ambient concentrations of pollutants on these values.

We are optimistic that the evolution of scientific knowledge and experience in applying the PSD provisions will result in a useful and workable program. The purposes that PSD is intended to serve are important and necessary. The coming months and years will provide the evidence necessary to judge the extent to which the existing provisions adequately serve these purposes.

APPENDIX: PERSONS CONSULTED IN THE STUDY

RICHARD ADAMS, University of Wyoming, Laramie
GALAN ANDERSEN, The Nokota Company, Bismarck, ND
LOWELL ASHBAUGH, University of California, Davis
SCOTT ATKINSON, American Petroleum Institute, Washington, DC
G.T. BADLEY, Montana Dakota Utilities, Bismarck, ND
WALTER BARBER, Environmental Protection Agency, Research Triangle Park, NC
DENNIS BATES, Metropolitan Council of Governments, Washington, DC
BRENT BRADFORD, Bureau of Air Quality, Salt Lake City, UT
KARL BRAITHWAITE, Los Alamos Scientific Laboratory, NM
DAVID BROOKSHIRE, University of Wyoming, Laramie
BARBARA BROWN, National Park Service, Washington, DC
STUART BURNESS, University of New Mexico, Albuquerque
DAVID BYRNES, Basin Electric Power Cooperative, Bismarck, ND
GENE CHRISTIANSON, North Dakota State Department of Health, Bismarck, ND
ROBERT COLLOM, Georgia Environmental Protection Division, Atlanta
JAY CRAWFORD, North Dakota State Department of Health, Bismarck, ND
KEN DEMERJIAN, Environmental Protection Agency, Research Triangle Park, NC
BRUCE EGAN, Environmental Research & Technology, Lexington, MA
ROBERT ERIKSEN, Basin Electric Power Cooperative, Bismarck, ND
RICHARD FERGUSON, SRI International, Menlo Park, CA

DAVID FOSTER, Environmental Protection Agency, Washington, DC
JAMES GALLOWAY, University of Virginia, Charlottesville
MARY ANN GRASSER, National Park Service, Denver, CO
CHARLES GOODMAN, Southern Services Company, Inc., Birmingham, AL
JEREMY HALES, Battelle Northwest, Richland, WA
DANIEL T. HEDDEN, Northeast Utilities Service Company
WALTER HERRIMAN, National Park Service, Chaco Canyon National Monument, NM
HARRY A. HOVEY, New York Division of Air Resources, Albany
T.L. HURST, Kerr-McGee Corporation, Oklahoma City, OK
WILL KAUL, Cooperative Power Association, Minneapolis, MN
E.R. KILPATRICK, Minnesota Power & Light, Duluth, MN
DOUGLAS LATIMER, Systems Application, Inc., San Rafael, CA
BRIAN LEADERER, Yale University School of Medicine, New Haven, CT
HOWARD LILJISTRAND, California Institute of Technology, Pasadena, CA
GEORGE LOU, Dames & Moore, Seattle, WA
EDWARD MACIAS, Washington University, St. Louis, MO
JAMES MAHONEY, Environmental Research & Technology, Concord, MA
VERLIN MENZE, Otter Tail Power Company, Fergus Falls, MN
JIM MILLER, Basin Electric Power Cooperative, Bismarck, ND
JOHN MOODY, Warren Petroleum Company, Tulsa, OK
DANA MOUNT, North Dakota State Department of Health, Bismarck, ND
BRIAN MURPHY, Environmental Research & Technology, Concord, MA
KEN McCONNELL, University of Maryland, College Park, MD
DAN McCONNON, United Power Association, Elk River, MN
DAVID NAKLES, Environmental Research & Technology, Pittsburgh, PA
HENRY NICKEL, Utility Air Regulatory Group, Washington, DC
JOHN NORTON, Dakota Resources Council, Dickinson, ND
RANDOLPH NUDLAND, Dunn County Citizens for Clean Air, Dunn Center, ND
GARY PARISH, Popham, Haik, Schnobrich, Kaufman & Doty, Ltd., Denver, CO
NEIL PAULSON, U.S. Forest Service, Washington, DC
MARK PITCHFORD, Environmental Protection Agency, Las Vegas, NV
RAFE POMERANCE, Friends of the Earth, Washington, DC
ALLAN RANDALL, University of Kentucky, Lexington
ROBERT REPETTO, Harvard School of Public Health, Cambridge, MA

AL RICKERS, Utah Bureau of Air Quality, Salt Lake City
MIKE ROACH, Montana Department of Health & Environmental Services, Helena
C.F. RUPE, Natural Gas Pipeline Company of America, Bismarck, ND
HAROLD SEITZ, New York Department of Environmental Conservation, Albany
EUGENE SESKIN, Department of Commerce, Washington, DC
JOHN SMITH, North Dakota Lignite Council, Bismarck, ND
DAVID STONEFIELD, Environmental Protection Agency, Research Triangle Park, NC
BRUCE TERRIS, Washington, DC
DAVID TORKELSON, North Dakota Department of Business and Industrial Development, Bismarck
MIKE TRUTNA, Environmental Protection Agency, Research Triangle Park, NC
HARVEY WICKWARE, National Park Service, Theodore Roosevelt National Park, ND
JIM WIEGOLD, Environmental Protection Agency, Research Triangle Park, NC
GARY WEINREICH, American Natural Service Company, Detroit, MI
WILLIAM WILSON, Environmental Protection Agency, Research Triangle Park, NC
BRION WISE, Western Gas Processors, Ltd., Northglenn, CO

GLOSSARY

Air quality model. A mathematical representation of the behavior of air pollutants or their effects on air quality related values.

Air quality related value (AQRV). A feature or property of an area--e.g., visibility--that is affected in some way by air pollution.

Attainment area. A region where ambient concentrations of criteria pollutants are lower than the levels specified as minimal standards for protecting public health and welfare, i.e., the NAAQS.

Baseline concentration. The concentration of a pollutant at the time of the first application for a PSD permit in an attainment area.

Best available control technology (BACT). An emission limitation applied to a new or modified major stationary source in an attainment area, determined on a case-by-case basis taking into account environmental consequences, energy considerations, and economic impacts.

Best available retrofit technology (BART). An emission limitation applied to an existing major stationary source that may reasonably be anticipated to cause or to contribute to impairment of visibility in a mandatory Class I areas.

<u>Class I, II, and III areas</u>. Regions in attainment areas where maintenance of existing good air quality is of high priority. In Class I areas, maintaining air quality is regarded as having the highest priority with respect to other values; in Class III areas, air quality has lower priority than it does in the other two areas. Initially, all attainment areas except mandatory Class I areas were designated Class II.

<u>Criteria pollutant</u>. A pollutant for which a national ambient air quality standard (NAAQS) has been established, i.e., sulfur dioxide, total suspended particulate mass, hydrocarbons, carbon monoxide, ozone, lead, and nitrogen dioxide.

<u>Federal land manager</u>. The Secretary of the department of the federal government under whose jurisdiction a piece of federal land is managed.

<u>Increments</u>. Maximum allowable increases over baseline concentrations of pollutants covered by the PSD provisions in Class I, II, and III areas. See Table 1.2 for values of the increments established by the Clean Air Act Amendments of 1977.

<u>Lowest achievable emission rate (LAER)</u>. An emission limitation applied to a new or modified major stationary source in a nonattainment area; the most stringent emission limitation contained in a state implementation plan or achieved in practice for a type of source.

<u>Major source</u>. Any source in an attainment area that emits at least 250 tons of any pollutant regulated under the Clean Air Act or any source from among 28 categories of sources that emits at least 100 tons per year of any regulated pollutant. See Table 1.3 for the 28 categories of sources.

<u>Mandatory Class I area</u>. An international park, a national wilderness area or national memorial park larger than 5,000 acres, or a national park larger than 6,000 acres. States may not reclassify mandatory Class I areas.

<u>Nonattainment area</u>. A region where ambient concentrations of criteria pollutants are higher than the

levels specified as minimal standards for protecting public health and welfare, i.e., the NAAQS.

Set I pollutants. Sulfur dioxide and total suspended particulate mass.

Set II pollutants. Criteria pollutants other than sulfur dioxide and total suspended particulate mass.

State implementation plan (SIP). A plan devised by a state and approved by the Administrator of the U.S. EPA for implementing and enforcing provisions of the Clean Air Act. Part of each SIP must describe a PSD program.

Total suspended particulate mass (TSP). A criteria pollutant measured as the mass of all particles in the atmosphere, without regard to size or chemical composition.

REFERENCES

Adams, D.F., S.O. Farwell, E. Robinson, and M.R. Pack (1980) Biogenic Sulfur Emissions in the SURE Region. EA-1516. Palo Alto, Cal.: Electric Power Research Institute.

Anderson, R.J., Jr., R.O. Reid, and E.P. Seskin (1979) An Analysis of Alternative Policies for Attaining and Maintaining a Short-Term NO_2 Standard. Princeton, N.J.: MATHTECH, Inc.

Angell, J.K. and J. Korshover (1975) Variation in sunshine duration over the continguous United States between 1950 and 1972. Journal of Applied Meteorology 14:1174-1181.

Arthur D. Little, Inc. (1980) The Effects of Prevention of Significant Deterioration on Industrial Development. Volume II of the Business Roundtable Air Quality Project. Washington, D.C.: The Business Roundtable.

Atkinson, S.E. and D.H. Lewis (1976) Determination and implementation of optimal air quality standards. Journal of Environmental Economics and Management 3:363-380.

Bass, A., D.G. Smith, G.F. Hoffnagle, and B.A. Egan (1978) Options to Improve and Upgrade Rough Terrain Models. P-3294-15. Concord, Mass.: Environmental Research and Technology.

Baumol, W.J. and W.E. Oates (1975) The Theory of Environmental Policy. Englewood Cliffs, N.J.: Prentice-Hall, Inc.

Bencala, K.E. and J.H. Seinfeld (1976) On frequency distributions of air pollutant concentrations. Atmospheric Environment 10:941-950.

Bolin, B. and R.J. Charlson (1976) On the role of the tropospheric sulfur cycle in the short-wave radiative climate of the earth. Ambio 5(2):47-54.

Bradford, D.F. (1970) Benefit-cost analysis and demand curves for public goods. Kyklos 23:775-791.

Brezonik, P.L., E.S. Edgerton, and C.D. Henry (1980) Acid precipitation and sulfate deposition in Florida. Science 208:1027-1028.

Brookshire, D.S., B.C. Ives, and W.D. Schulze (1976) The valuations of aesthetic preferences. Journal of Environmental Economics and Management 4:325-346.

Brown, H.D., P.R. Miller, J.M. Skelly, D.B. Drummond, and C.E. Carlson (1979) Air Pollution Effects on Forest Vegetation and the Analysis of the Role of Forest Insect and Disease Management. Washington, D.C.: Forest Service, U.S. Department of Agriculture.

Buhyoff, G.J. and W.A. Leuschner (1978) Estimating psychological disutility from damaged forest stands. Forest Science 24:231-246.

Butler, R.L., E.L. Cooper, J. K. Crawford, D.C. Hales, and W.G. Kimmel (1973) Fish and Food Organisms in Acid Mine Waters of Pennsylvania. EPI-23-73-032. Washington, D.C.: U.S. Government Printing Office.

California Air Resources Board (undated) An Introduction to the California Air Conservation Program and the Federal Prevention of Significant Deterioration Program. Sacramento, Cal.

Cass, G.R. (1976) The Relationship between Sulfate Air Quality and Visibility at Los Angeles. Caltech Environmental Quality Laboratory Memorandum No. 18. Pasadena, Cal.

Changnon, S.A., Jr. (1968) The La Porte weather anomaly: fact or fiction? Bulletin of the American Meteorology Society 49:4-11.

Charlson, R.J., A.P. Waggoner, and J.F Thiekle (1978) Visibility Protection for Class I Areas: The Technical Basis. NTIS PB-288 842. Springfield, Va.: National Technical Information Service.

Cooper, J.A. and J.G. Watson (1979) Portland Aerosol Characterization Study (PACS). Final Report Summary, prepared for Portland Air Quality Maintenance Area Advisory Committee and Oregon Department of Environmental Quality.

Countess, R.J., G.T. Wolff, and S.H. Cadle (1980) Denver aerosol: a comprehensive chemical characterization. _In_ Proceedings of the 73rd Annual Meeting of the Air Pollution Control Association, Montreal, Quebec. Pittsburgh, Pa.: Air Pollution Control Association.

Courtney, W.J., J.W. Tesch, R.K. Stevens, and T.G. Dzubay (1980) Characterization of Denver aerosol between December 1978 and December 1979. In Proceedings of the 73rd Annual Meeting of the Air Pollution Control Association, Montreal, Quebec. Pittsburgh, Pa.: Air Pollution Control Association.

Crocker, T.D. (1966) Structuring of atmospheric pollution control systems. In The Economics of Air Pollution, edited by H. Wolozin, pp. 61-86. New York, N.Y.: W.W. Norton.

Crocker, T.D. (1973) Contractual choice. National Resources Journal 13:561-577.

Crocker, T.D., J.T. Tschirhart, R.M. Adams, and B. Forster (1980) Methods Developed for Assessing Acid Precipitation Control Benefits. Draft report prepared for U.S. Environmental Protection Agency. Laramie, Wyo.: University of Wyoming.

Dales, J.H. (1968) Pollution, Property, and Prices. Toronto, Can.: University of Toronto Press.

Dames and Moore (1980) Study of PSD and Offset Permits. Prepared for the National Commission on Air Quality, contract no. NCAQ 1-AQ-7133. Seattle, Wash.: Dames and Moore.

Dana, M.T., J.M. Hales, and M.A. Wolf (1975) Rain scavenging of SO_2 and sulfate from power plant plumes. Journal of Geophysical Research 80:30.

Davis, R. (1963) Recreation planning as an economic problem. Natural Resources Journal 3:239-249.

Dzubay, T.G. (1979) Chemical element balance method applied to dichotomous sampler data. In Proceedings of the Conference on Aerosols: Anthropogenic and Natural Sources and Transport. New York: New York Academy of Sciences.

Eliassen, A. (1980) A review of long-range transport modeling. Journal of Applied Meteorology 19(3):231-240.

Environmental Research and Technology (1978) Draft, Phase I Report, An Analysis of Clean Air Act Provisions to Protect Visibility. Document P-3294-19. Palo Alto, Cal.

Eriksen, R.L. (1980) The Development of Dry Flue Gas Desulfurization at Basin Electric Power Cooperative. Presented at the Second Conference on Air Quality Management in the Electric Power Industry, Austin, Tex., January 22-25.

Eriksson, E. (1960) The yearly circulation of chloride and sulfur in nature; meteorological, geochemical and pedological implications. Part II. Tellus 12:63-109.

Flowers, E.C., R.A. McCormick, and K.R. Kurfis (1969) Atmospheric turbidity over the United States, 1961-1966. Journal of Applied Meteorology 8:955-962.

GCA Corporation (1980) Acid Rain Information Book. GCA TR-80-87-G. Draft final report prepared for the U.S. Department of Energy. Bedford, Mass.

Ginevan, M.E., D.D. Lane, and L. Greenberg (1980) Ambient air concentrations of sulfur dioxide affects flight activity in bees. Proceedings of the National Academy of Sciences 77(10):5631-5633.

Hales, J.M. (1972) Fundamentals of the theory of gas scavenging by rain. Atmospheric Environment 6:635-659.

Hales, J.M., M.A. Wolf, and M.T. Dana (1973) A linear model for predicting the washout of pollutant gases from industrial plumes. AIChE Journal 19(2):292-297.

Hales, J.M. and M.T. Dana (1979a) Precipitation scavenging of urban pollutants by convection storm systems. Journal of Applied Meteorology 8:3.

Hales, J.M. and M.T. Dana (1979b) Regional-scale deposition of sulfur dioxide by precipitation scavenging. Atmospheric Environment 13:1121-1132.

Hanna, S.R., G.A. Briggs, J. Deardorff, B.A. Egan, F.A. Gifford, and F. Pasquill (1977) AMS workshop on stability classification schemes and sigma curves: summary of recommendations. Bulletin of American Meteorological Society 58(12):1305-1309.

Harward, M. and M. Treshow (1975) Impact of ozone on the growth and reproduction of understory plants in the aspen zone of western U.S.A. Environmental Conservation 2(1):17-23.

Heisler, S.L., R.C. Henry, and J. Collins (1980) The Nature of the Denver Haze in November and December 1978. In Proceedings of the 73rd Annual Meeting of the Air Pollution Control Association, Montreal, Quebec. Pittsburgh, Pa.: Air Pollution Control Association.

Henderson, R.G., R. Pitter, and J. Wisniewski (1980) Research Guidelines for Regional Modeling of Fine Particulates, Acid Deposition and Visibility. MTR-80W148. McLean, Va.: The MITRE Corporation.

Holzworth, G.C. (1962) Some Effects of Air Pollution on Visibility in and near Cities. SEC Technical Report A62-5. Washington, D.C.: U.S. Government Printing Office.

Huff, E.A. and S.A. Changnon (1973) Precipitation modification by major urban areas. Bulletin of the American Meteorological Society 54:1220-1232.

Husar, R.B., N.V. Gillana, and J.P. Husar (1976) A Study of Long Range Transport from Visibility Observations, Trajectory Analysis and Local Air Pollution Monitoring Data. Presented at the 7th International Technical Meeting on Air Pollution Modeling and its Application. Airlie, Va.: NATO/CCMS.

Husar, R.B., D.E. Patterson, and J. Holloway (1979) Trends of the Eastern U.S. Haziness since 1948. Presented at the Fourth Symposium on Atmospheric Turbulence, Diffusion, and Air Pollution of the American Meteorological Society, Reno, Nev.

ICF, Inc. (1980) Preliminary Assessment of Economic Impact of Visibility Regulations. Washington, D.C.

Illinois Environmental Protection Agency (1980) A Study to Determine the Effects of Reclassifying Portions of Madison, Monroe and St. Clair Counties from Class II To Class III. Springfield, Ill.

Johnson, C.D., D.A. Latimer, R.W. Bergstrom, and H. Hogo (1980) User's Manual for the Plume Visibility Model (PLUVUE). EPA-450/5-80-032. Springfield, Va.: National Technical Information Service.

Keith R.W. (1964) A Study of Low Visibilities in the Los Angeles Basin, 1950-1961. Air Quality Report #65. Los Angeles, Cal.: Los Angeles Air Pollution Control District.

Keith, R.W. (1970) Downtown Los Angeles noon visibility trends, 1933-1969. Air Quality Report #65. Los Angeles, Cal.: Los Angeles Air Pollution Control District.

Kress, L.W., J.M. Skelly, and R.G. Wilhour (1980) Prevention of Significant Deterioration of Native Ecosystems: Allowable Pollutant Concentrations. Corvallis, Ore.: U.S. Environmental Protection Agency.

Landsberg, H.E. (1956) The climate of towns. In Man's Role in Changing the Face of the Earth, edited by W.L. Thomas, Jr., pp. 584-606. Chicago, Ill.: University of Chicago Press.

Larsen, R.I. (1971) A Mathematical Model for Relating Air Quality Measurements to Air Quality Standards. EPA-AP-89. Research Triangle Park, N.C.: U.S. Environmental Protection Agency.

Latimer, D.A. (1979) Power Plant Impacts on Air Quality and Visibility: Siting and Emission Control Implications. EF79-101. San Rafael, Cal.: Systems Applications, Inc.

Latimer, D.A., R.W. Bergstrom, S.R. Hayes, M. Liu, J.H. Seinfeld, G.Z. Whitten, M.A. Wojcik, and M.J. Hillyer (1978) The Development of Mathematical Models for the Prediction of Anthropogenic Visibility Impairment, vol. I. EPA 450/3-78-110a. Research Triangle Park, N.C.: U.S. Environmental Protection Agency.

Latimer, D.A. and R.G. Ireson (1980) Workbook for Estimating Visibility Impairment. EPA 450/4-80-031. Springfield, Va.: National Technical Information Service.

Leaderer, B.P. and J.A. Stolwijk (1979) Optical properties of urban aerosols and their relation to chemical composition. In Proceedings of the Conference on Aerosols: Anthropogenic and Natural Sources of Transport, pp. 70-85. New York: New York Academy of Science.

LeBlanc, F. and D.N. Rao (1975) Effects of Air Pollution on Lichens and Bryophytes. In Responses of Plants to Air Pollution, ed. J.B. Mudd and T.T. Kozlowski. New York: Academic Press.

Legge, A.H., D.R. Jaques, G.W. Harvey, H.R. Krause, H.M. Brown, E.D. Rhodes, M. Nosal, H.W. Schellhase, J. Mayo, A.P. Hartgerink, P.F. Lester, R.G. Amundson, and R.B. Walker (1978) Sulfur Gas Emissions in the Boreal Forest: The West Whitecourt Case Study. Kananaskis Center for Environmental Research. Alberta, Canada: University of Calgary.

Leonardos, G., D. Kindall, and N. Barnard (1969) Odor threshold determinations of 53 odorant chemicals. Journal of the Air Pollution Control Association 19(2):91-95.

Lewis, C.W. and E.S. Macias (1980) Composition of size-fractionated aerosol in Charleston, West Virginia. Atmospheric Environment 14:185-194.

Lewis, W.M., Jr. and M.C. Grant (1980) Acid precipitation in the western United States. Science 207:176-177.

Likens, G.E., R.F. Wright, J.N. Galloway, and T.J. Butler (1979) Acid rain. Scientific American 241:43-51.

Liljestrand, H.M. and J.J. Morgan (1978) Chemical composition of acid precipitation in Pasadena, California. Environmental Science & Technology 12:12.

Loehman, E., D. Boldt, and K. Chaiken (1980) Measuring the Benefits of Air Quality Improvements in the San Francisco Bay Area. Draft report for U.S. Environmental Protection Agency. Menlo Park, Cal.: SRI International.

Lyons, C.E., I. Tombach, R.A. Eldred, F.P. Terraglio, and J.E. Core (1979) Relating Particulate Matter Sources and Impacts in the Willamette Valley During Field and Slash Burning. Presented at the Annual Meeting of the Air Pollution Control Association, Cinncinati, Ohio. Pittsburgh, Pa.: Air Pollution Control Association.

Macias, E.S., D.L. Blumenthal, J.A. Anderson, and B.K. Cantrell (1979) Characterization of Visibility-Reducing Aerosols in the Southwestern United States. Interim Report on Project VISTTA. Meteorology Research, Inc. MRI 78IR-1585. Washington, D.C.: U.S. Environmental Protection Agency.

Macias, E.S., J.D. Zwicker, J.R. Ouimette, S.V. Hering, S.K. Friedlander, T.A. Cahill, G.A. Kuhlmey, L.W. Richards, and W.H. White (1980) Regional Haze in the Southwestern U.S.: Aerosol Chemical Composition. Presented at the Symposium on Visibility: Measurements and Model Components, Grand Canyon, Ari., November, 1980.

Malm, W., K.K. Leiker, and J.V. Molenar (1979a) Human Perception of Visual Air Quality. <u>In</u> Proceedings: View on Visibility--Regulatory and Scientific. Rocky Mountain States Section, Air Pollution Control Association. Pittsburgh, Pa.: Air Pollution Control Association.

Malm, W., K.K. Leiker, and J.V. Molenar (1979b) Visibility in the Southwest, Environmental Monitoring Systems Laboratory, U.S. Environmental Protection Agency. Las Vegas, Nev.: U.S. Environmental Protection Agency.

Marians, M. and J. Trijonis (1979) Empirical Studies of the Relationship between Emissions and Visibility in the Southwest. Prepared by Technology Service Corporation. Santa Fe, N.M.: U.S. Environmental Protection Agency.

Marin, A. (1978) The choice of efficient pollution policies: technology and economics in the control of sulfur dioxide. Journal of Environmental Economics and Management 5:44-62.

McCormick, R.A. and J.H. Ludwig (1967) Climate modification by atmospheric aerosols. Science 156:1358-1360.

Miller, P.R., editor (1980) Proceedings of Symposium on Effects of Air Pollutants on Mediterranean and Temperate Forest Ecosystems, Riverside, Cal., June 22-27, 1980. Berkeley, Cal.: Pacific Southwest Forest and Range Experiment Station.

MITRE Corporation (1979) National Environmental Impact Projection No. 1. HCP/P-6119. McLean, Va.

Muskie, E. (1977) Remarks printed in Congressional Record, June 8.

National Research Council (1977a) Nitrogen Oxides. Washington, D.C.: National Academy of Sciences.

National Research Council (1977b) Ozone and Other Photochemical Oxidants. Washington, D.C.: National Academy of Sciences.

National Research Council (1977c) Environmental Monitoring. Washington, D.C.: National Academy of Sciences.

National Research Council (1978) Sulfur Oxides. Washington, D.C.: National Academy of Sciences.

National Research Council (1979a) Hydrogen Sulfide. Baltimore, Md.: University Park Press.

National Research Council (1979b) Ammonia. Baltimore, Md.: University Park Press.

National Research Council (1979c) Odors from Stationary and Mobile Sources. Washington, D.C.: National Academy of Sciences.

National Research Council (1979d) Airborne Particles. Baltimore, Md.: University Park Press.

National Research Council (1979e) Carbon Dioxide and Climate: A Scientific Assessment. Washington, D.C.: National Academy of Sciences.

National Research Council (1980) Controlling Airborne Particles. Washington, D.C.: National Academy of Sciences.

Newman, J.R. (1980a) Effects of Air Emissions on Wildlife Resources. FWS/OBS-80/40.1. Ann Arbor, Mich.: Fish and Wildlife Service, U.S. Department of Interior.

Newman, L. (1980b) Atmospheric Oxidation of Sulfur Dioxide as Viewed from Power Plant and Smelter Plume Studies. Presented at Symposium on Plumes and Visibility: Measurements and Model Components, Grand Canyon, Ariz., November, 1980.

Ogren, J.A., D.L. Blumenthal, and A.H. Vanderpol (1976) Oxidant Measurements in Western Power Plant Plumes. Prepared by Meteorology Research Inc. Palo Alto, Cal.: Electric Power Research Institute.

Pack, D.H. (1980) Precipitation chemistry patterns: a two-network data set. Science 280:1143-1145.

Perera, F.P. and A.K. Ahmed (1978) Respirable Particles. Washington, D.C.: Natural Resources Defense Council, Inc.

Peterson, J.T. and E.C. Flowers (1977) Interactions between air pollution and solar radiation. Solar Energy 19:23-32.

Raddum, G. (1978) Invertebrates: Quality and quantity as fish food. In Limnological Aspects of Acid Precipitation, edited by G. Hendry, pp. 21-42. Upton, N.Y.: Brookhaven National Laboratory.

Raffle, B.I. (1979) Prevention of Significant Deterioration and Nonattainment under the Clear Air Act: A Comprehensive Review. Environmental Reporter, Monograph No. 27, 10(1). Washington, D.C.: Bureau of National Affairs.

Randall, A.S., B.C. Ives, and E. Eastmen (1974) Bidding games for valuation of aesthetic environmental improvements. Journal of Environmental Economics and Management 1:132-149.

Repetto, R. (1980) The Influence of Standards, Effluent Charges and Other Regulatory Approaches on Innovation in Abatement Technology. Working Paper, Center for Population Studies, Harvard University, Cambridge, Mass.

Rose-Ackerman, S. (1973) Effluent charges: a critique. Canadian Journal of Economics 6:210-231.

Rowe, R.D., R.C. d'Arge, and D.S. Brookshire (1980) An experiment on the economic value of visibility. Journal of Environmental Economics and Management 7:1-19

Schulze, W.D. , D.S. Brookshire, E. Walther, and K. Kelly (1980) The Value of Visibility in the National Parklands of the Southwest. Draft report for U.S. Environmental Protection Agency.

Scott, B.C. (1978) Parameterization of sulfate removal by precipitation. Journal of Applied Meteorology 17:9.

Semonin, R.G. (1974) Urban-induced weather modification. In Earth and Environment and Resources Conference Digest of Technical Papers, edited by Lewis Winner, pp. 40-41. New York.

STAPPA (1980) Recommended Changes to the Clean Air Act. Washington, D.C.: State and Territorial Air Pollution Program Administrators.

Tombach, I. (1978) A Critical Review of Methods for Defining Visual Range in Pristine and Near-Pristine Areas. Presented at the 71st Annual Meeting of the Air Pollution Control Association, Houston, Tex. Pittsburgh, Pa.: Air Pollution Control Association.

Tombach, I. (1979) Intercomparison of visibility measurement methods. In Proceedings: View on

Visibility--Regulatory and Scientific. Rocky Mountain States Section, Air Pollution Control Association. Pittsburgh, Pa.: Air Pollution Control Association.

Trijonis, J. (1979) Visibility in the Southwest: an exploration of the historical data base. Atmospheric Environment 13:833-843.

Trijonis, J. (1980) Visibility in California. Prepared by Technology Service Corporation for California Air Resources Board. Contract #A7-181-30. Springfield, Va.: National Technical Information Service.

Trijonis, J., J. Eldon, J. Gins, and G. Berglund (1980) Analysis of the St. Louis RAMS Ambient Particulate Data. EPA-450/80-006a. Research Triangle Park, N.C.: U.S. Environmental Protection Agency.

Trijonis, J. and D. Shapland (1979) Existing Visibility Levels in the U.S.: Isopleth Maps of Visibility in Suburban/Nonurban Areas during 1974-1976. EPA-450/5-79-010. Research Triangle Park, N.C.: U.S. Environmental Protection Agency.

Trijonis, J. and K. Yuan (1978a) Visibility in the Southwest: An Exploration of the Historical Data Base. EPA-600/3-78-039. Washington, D.C.: U.S. Environmental Protection Agency.

Trijonis, J. and K. Yuan (1978b) Visibility in the Northeast: Long-Term Visibility Trends and Visibility/Pollutant Relationships. EPA-600/3-78-075. Washington, D.C.: U.S. Environmental Protection Agency.

Treshow, M. and J. Allan (1979) Annual variation in the dynamics of a woodland plant community. Environmental Conservation 6(3):231-236.

U.S. Environmental Protection Agency (1976) National Emissions Data System (NEDS) of the Aerometric and Emissions Reporting System (AEROS). EPA 450/2-76/007. Research Triangle Park, N.C.

U.S. Environmental Protection Agency (1978) Guideline on Air Quality Models. EPA-450/2-78-027. Research Triangle Park, N.C.

U.S. Environmental Protection Agency (1979a) Protecting Visibility: An EPA Report to Congress. EPA-450/5-79-008. Research Triangle Park, N.C.

U.S. Environmental Protection Agency (1979b) Research Summary: Acid Rain. EPA-600/8-79-028. Washington, D.C.

U.S. Environmental Protection Agency (1979c) Compilation of BACT/LAER Determinations. EPA-450/2-79-003. Research Triangle Park, N.C.

U.S. Environmental Protection Agency (1979d) Guidance for Lowest Achievable Emission Rates from 18 Major Stationary Sources of Particulate, Nitrogen Oxides, Sulfur Dioxide, or Volatile Organic Compounds. EPA-450/3-79-024. Research Triangle Park, N.C.

U.S. Environmental Protection Agency (1980a) Proposed Guidelines for Determining Best Available Retrofit Technology for Coal-Fired Power Plants and Other Major Stationary Sources. EPA-450/3-80-009a. Research Triangle Park, N.C.

U.S. Environmental Protection Agency (1980b) Control Techniques for Particulate Emissions from Stationary Sources. Preliminary draft in two volumes. Research Triangle Park, N.C.

U.S. Environmental Protection Agency (1980c) Proposed Revisions, Guidelines on Air Quality Models. Research Triangle Park, N.C.

U.S. House of Representatives (1977) Report of the Interstate and Foreign Commerce Committee to Accompany H.R. 6161, no. 95-294. Reprinted in U.S. Code Congressional and Administrative News, 95th Congress, no. 8. St. Paul, Minn.: West Publishing Company.

U.S. Senate (1977) Report of the Committee on Environment and Public Works to accompany S. 252, no. 95-127. 95th Congress. Washington, D.C.: U.S. Government Printing Office.

White, W.H. and P.T. Roberts (1977) On the nature and origins of visibility-reducing aerosols in the Los Angeles air basin. Atmospheric Environment 11:803.

WMO/EPA/NOAA/UNEP (1977) Global Monitoring of the Environment for Selected Atmospheric Constituents, 1975. Asheville, N.C.: Environmental Data Service, National Climatic Center.

Wolff, G.T., R.J. Countess, P.J. Groblicki, M.A. Ferman, S.H. Cadle, and J.L. Muhlbaier (1980) Visibility-Reducing Species in the Denver "Brown Cloud," Part II, Sources and Temporal Patterns. Presented at the Symposium on Visibility: Measurements and Model Components, Grand Canyon, Nev., November 1980.

World Health Organization (1972) Air Quality Criteria and Guides for Urban Air Pollutants. No. 506. Geneva, Switzerland.

Yohe, G.W. (1977) Comparisons of price and quantity controls: a survey. Journal of Comparative Economics 1:213-233.